Molecular Diagnostics

E. van Pelt-Verkuil · W.B. van Leeuwen
R. te Witt
Editors

Molecular Diagnostics

Part 2: Clinical, Veterinary, Agrobotanical and Food Safety Applications

Springer

Editors
E. van Pelt-Verkuil
Faculty of Science and Technology
University of Applied Sciences Leiden
Leiden
The Netherlands

R. te Witt
Netherlands Molecular Diagnostic Laboratory (NMDL)
Rijswijk
The Netherlands

W.B. van Leeuwen
Faculty of Science and Technology
University of Applied Sciences Leiden
Leiden
The Netherlands

ISBN 978-981-10-4510-3 ISBN 978-981-10-4511-0 (eBook)
DOI 10.1007/978-981-10-4511-0

Library of Congress Control Number: 2017937518

Printed on acid-free paper

This Springer imprint is published by Springer Nature
The registered company is Springer Nature Singapore Pte Ltd.
The registered company address is: 152 Beach Road, #21-01/04 Gateway East, Singapore 189721, Singapore

Contents

Introduction

In a very short historical time span of only 30 years, molecular diagnostic strategies have been efficiently embedded in a wide variety of both clinical and veterinary and agro botanical disciplines.

In 2001 E. van Pelt-Verkuil, M.F. van Berlo, A.F. van Belkum and H.H.M. Niesters stood at the cradle of a first book on molecular diagnostics.

In 2013, the second printing of "Molecular Diagnostics", was published in the Dutch language: "Moleculaire Diagnostiek". For this book print, a new editorial team was formed, consisting of W.B. van Leeuwen and E. van Pelt-Verkuil. Immediately after publication it seemed that there was a great need for an English translation of this book.

For this translation and the third print, R. te Witt was added to the last editorial team. Now, in 2017, the third print lies before you. The editors decided to publish the scientific content of the book in two volumes. Volume 1 describes the theoretical background and Volume 2 addresses the implementation and application of molecular methods. Furthermore, the book was fully updated to the current standards and guidelines. All chapters are written by experts in the different concerning disciplines.

Both volumes focus on the implementation of molecular techniques for routine use in both clinical and research laboratories. The first volume discusses a substantial compilation of collective knowledge regarding validation, implementation, evaluation and quality control of currently available molecular techniques. These techniques include not only PCR technology but also isothermal amplification methods, high-throughput sequencing and bioinformatics. This volume concludes with a chapter on MALDI-TOF with a special focus on single cell MALDI-TOF.

In the second volume, many different disciplines using molecular diagnostics are well presented. Multiple specialists describe different applications of these technologies in their own disciplines, such as clinical medicine (parasitology, virology, bacteriology, mycology, hematology, pathology, pharmacogenetics and clinical genetics), plant pathology, food safety and veterinary medicine.

The editors thank all authors for their professional effort in the development and realization of the two volumes.

Chapter 1
Molecular Diagnostics of Bacterial Pathogens

Petra F.G. Wolffs, Tanja H. Geelen and Lieke B. van Alphen

Abstract Molecular diagnostics of bacterial pathogens is a rapidly growing field within clinical microbiology. The first commonly used routine application was the detection of bacterial sexually transmitted pathogens. Most of these micro-organisms are difficult to culture and therefore molecular diagnostics are providing a reliable, easy and high-throughput alternative. Over time, molecular diagnostic methods were developed and implemented for more easily culturable micro-organisms as well. The implementation of molecular tests for detection of bacterial gastrointestinal pathogens is one example that is discussed in this chapter. In this latter area often a combination of molecular diagnostics with conventional culture is preferred. In the final part of the chapter, bacterial typing is highlighted. Molecular typing of bacteria has provided the opportunity, in contrast to conventional typing, to perform same day, high throughput typing. Furthermore, many of the molecular typing methods discussed in this chapter have an improved discriminatory power. In conclusion, the field of molecular diagnostics of bacterial pathogens is constantly moving and new techniques are being developed and implemented in a continuous search for assays with an even higher technical and diagnostic sensitivity and/or specificity. The importance of molecular diagnostics of bacterial pathogens for clinical microbiology laboratories will likely only further increase in the future.

Keywords Bacterial pathogens · Sexually transmitted disease · Gastroenteritis · Antimicrobial treatment · Molecular typing

1.1 Introduction

This chapter aims to give a non-comprehensive overview of molecular diagnostics of clinically relevant bacterial pathogens. Molecular diagnostics of bacterial pathogens from clinical samples is a relatively new, but rapidly growing field. In contrast to the

P.F.G. Wolffs (✉) · T.H. Geelen · L.B. van Alphen
Department of Medical Microbiology, Maastricht UMC, Maastricht, The Netherlands
e-mail: p.wolffs@mumc.nl

E. van Pelt-Verkuil et al. (eds.), *Molecular Diagnostics*,
DOI 10.1007/978-981-10-4511-0_1

difficult-to-culture, slow-growing viruses, for which molecular diagnostics was rapidly developed to overcome those diagnostic challenges, for bacteria, the ease and low costs of culture-based diagnostics initially limited the application of molecular diagnostics. However, in recent years, also for many bacterial pathogens, molecular tests have been developed. In this chapter we will illustrate developments in this field using two examples. First we will discuss molecular assays for the detection of bacterial sexually transmitted infections. Molecular bacterial diagnostics started in most clinical laboratories in this area. The second example concerns the diagnosis of bacterial pathogens involved in gastroenteritis. In contrast, this is an area where developments are more recent, but are widely being implemented. At the end of the chapter we will focus on the use of molecular techniques for typing of bacteria by briefly discussing the (choice of) techniques and their added value in a clinical laboratory that is investigating bacterial pathogens.

1.2 Molecular Diagnostics of Bacterial Sexually Transmitted Infections

1.2.1 Molecular Diagnostics of Chlamydia trachomatis and Neisseria gonorrhoeae

Globally, more than 1 million people acquire a sexually transmitted infection every day. The most prevalent bacterial pathogens that cause sexually transmitted infections (STIs) are *Chlamydia trachomatis* and *Neisseria gonorrhoeae*. The most recent WHO estimates indicated that in 2008, 106 million new cases of both *C. trachomatis* and *N. gonorrhoeae* occurred worldwide and that these incidences were significantly increased from those in 2005. Conventional diagnostics of these STIs is technically challenging, as *C. trachomatis* is a strict intracellular pathogen and requires insensitive and time-consuming tissue culturing to facilitate growth, while *N. gonorrhoeae* culture has demanding nutritional and environmental growth requirements. As a consequence, the use of molecular diagnostics for the detection of both pathogens has long since been the method of choice and is recommended by for instance the FDA. In addition to the possibilities for fast, high-throughput detection of these pathogens by molecular diagnostics and the increased sensitivity as compared to culture, additional advantages include the suitability of self-sampling kits for these tests and the less stringent transport conditions required for samples.

Due to the preference for molecular diagnostics for detection of *C. trachomatis* and *N. gonorrhoeae* and the high quantity of testing performed, these were among the first targets for which several commercially available molecular assays were available. Commercially available tests used (real-time) PCR-based amplification and detection, but also (real-time) transcription mediated amplification (TMA) and strand displacement amplification (SDA). In brief, TMA is an RNA-mediated isothermal amplification reaction, using both reverse transcriptase and RNA polymerase to generate billions of RNA amplicons. The process is initiated by the hybridisation of a primer that includes a RNA polymerase promotor sequence to an RNA template, followed by a reverse transcription reaction during which cDNA is formed. After denaturation of the initial RNA strand, a complementary DNA strand is formed. Both strands of the resulting double-stranded DNA molecule (including an RNA polymerase promotor sequence in both strands) are then used as templates for transcription and transcribed multiple times. The formed RNA molecules are also themselves again templates for further amplification. The TMA process strongly resembles the more commonly known nucleic acid sequence based amplification (NASBA), in which different enzymes are used to generate a comparable reaction. In contrast, SDA consists of an isothermal amplification of DNA, where a strand-displacing enzyme, such as *Bst* DNA Polymerase or Klenow Fragment, initiates replication at nicks at a single strand of the double-stranded DNA, by using a so called strand-limited restriction endonuclease (typically *HincII*). The nicking site is part of the primer and is regenerated during each amplification event, after which nicking can take place again and so a cascade of amplification can take place.

Independent of the amplification technique used, most commercially available assays have shown a comparable performance in independent external quality assurance trials with perhaps a slightly higher analytical sensitivity for some of the RNA-based amplification methods. Overall, the main developments in recent years have been that many commercial assays are incorporating a dual-target approach. The importance of this approach was highlighted by the discovery of new variants of especially *C. trachomatis*, such as the so-called Swedish variant which showed a 377 bp deletion in the *C. trachomatis* plasmid, the deleted fragment contains a highly popular target site used for different amplification assays. Further developments include increasing automation of the assays to facilitate high-throughput analysis, and a shift to assays with an even higher analytical sensitivity.

The main purpose of the diagnostics of *C. trachomatis* and *N. gonorrhoeae* is to guide treatment, although treatment might also be initiated based on symptoms and prior to diagnostic results. However, symptoms (often self-reported) have shown to be of limited value in predicting positivity of samples and patients. Furthermore, many *C. trachomatis* and *N. gonorrhoeae* infections are asymptomatic. For this reason, there is a high interest in speeding up the diagnostics process and where possible, to develop point-of-care tests. Current point-of-care assays, such as rapid antigen tests, have thus far lacked sensitivity and/or specificity and are of limited use.

Aside from speeding up the time from sample-to-result, an additional challenge in the diagnostics of these pathogens is that hardly any available assays have been validated or approved for use on extra-genital samples. An important hurdle in the validation has been the lack of routine testing at these sites, resulting in limited data and samples for validation. Furthermore, for *N. gonorrhoeae*, the presence of high concentrations of commensal *Neisseria* species has sometimes led to false-positive PCR results. This has also resulted in hesitation to use molecular diagnostics for detection at extra-genital sites. However, the importance of anorectal and oropharyngeal sexually transmitted bacterial pathogens has been highlighted in recent years. Recent research has shown that up to 75% of *C. trachomatis* and *N. gonorrhoeae* infections are missed when extra-genital samples are not tested and it is now believed that this positivity potentially plays a significant role in the transmission of these pathogens. Nonetheless, no routine testing at these sites is currently recommended, mostly due to the costs involved and testing of these sites is performed on request or for specific high-risk groups such as men who have sex with men only.

Finally, an important challenge specifically addressing *N. gonorrhoeae* is the rising global phenomenon of antibiotic resistance of this micro-organism against most therapeutic options. The current gold-standard technique for determining an antibiotic susceptibility pattern involves a culture of the bacterium in the presence of increasing concentrations of the antibiotic. However, as mentioned before, culturing *N. gonorrhoeae* is difficult and unsuccessful in up to 50% of PCR-positive patients. Monitoring the spread of potentially resistant strains is therefore severely limited. Future research will have to show whether reliable diagnostic molecular tests can be developed to monitor the presence of antibiotic resistance without the need for culturing. This would have to take place by detection of the antibiotic resistance genes (or mutations). While this technique has been used in clinical laboratories for important antibiotic resistant pathogens, such as methicillin resistant *Staphylococcus aureus* (MRSA) or vancomycin resistant enterococci (VRE), for *N. gonorrhoeae*, the mechanisms of antibiotic resistance seem to be complex and not fully understood.

1.2.2 Molecular Diagnostics of Other Sexually Transmitted Bacteria

Although *C. trachomatis* and *N. gonorrhoeae* are the most frequently found sexually transmitted bacterial diseases, in recent years other bacteria are also gaining interest. Especially for *Mycoplasma genitalium*, it is becoming clear from transmission studies involving bacterial typing (discussed later in this chapter) that these bacteria are spread from one person to a sex partner. This bacterium is mostly involved in non-gonococcal urethritis (painful or difficult urination) in men. In women, it can also cause similar symptoms as well as infections of the cervix. Like *C. trachomatis* and even more than *N. gonorrhoeae, M. genitalium* is difficult to culture. It requires inoculation in tissue cultures and even then can take weeks to grow. Using molecular diagnostics is therefore the only realistic means of detection. In contrast to *C. trachomatis* and *N. gonorrhoeae*, there are hardly any validated commercial assays available for *M. genitalium* and there is no consensus on who should be screened or tested for this pathogen. Nonetheless, recent reports investigating the prevalence of *M. genitalium* have shown increasing rates. In Denmark, a nationwide survey in a risk-population showed detection rates of up to 4% in woman and up to 10% in men. More importantly, antibiotic resistance seems highly common, with 40% of the positive samples in the Danish study being resistant to macrolide antibiotics (a common treatment for non-gonococcal urethritis). Due to difficulties in culturing of this bacterium, assessment of antibiotic resistance has to occur through molecular diagnostics. However, this is currently only performed in highly specialised laboratories.

Finally, other bacteria that are potentially involved in sexually transmitted infections are increasingly being investigated by molecular diagnostics. Similar to the situation with *Mycoplasma genitalium*, for *Ureaplasma urealyticum*, which also causes urethritis, there are hardly any commercial assays available and non-routine diagnosis is performed with in-house PCR protocols. There is growing evidence that imbalance in the vaginal microbiota increases susceptibility to different sexually transmitted infections. There are different molecular methods available (such as next generation sequencing and quantitative PCR algorithms) that are used to investigate changes in the vaginal microbiota.

1.3 Molecular Diagnostics of Bacterial Gastroenteritis

The second example discussed in this chapter regards bacterial gastroenteritis, which remains an enormous global health problem. The burden of disease is most critical in developing countries, especially among children under 5 years of age. It is also an important cause of morbidity and is associated with significant health-care costs in high-income countries. Here, bacterial gastroenteritis in usually self-limiting and, most of the time, neither empirical antimicrobial therapy nor laboratory diagnostics are needed. Prevention of transmission is the most important measure, especially in patients with acute diarrhoea. In contrast, identification of an etiological agent is required for management of patients with severe diarrhoea, for those showing symptoms that are consistent with invasive disease or with complications. In this case, clinical diagnosis guides clinicians in making the decision to start appropriate therapy as early as possible. Besides the value for individual patient care, characterization of bacterial enteropathogens is also necessary for hospitals to implement correct infection control measures and for public health officials to identify and track outbreaks of bacterial gastroenteritis.

1.3.1 Major Bacterial Pathogens for Routine Microbiological Diagnostics

Diarrhoea is usually a symptom of an infection in the intestinal tract, which can be caused by a variety of bacteria, as well by viruses or parasites. However, the aetiology of diarrhoea can also be non-infectious, caused by food intolerances, reactions to medications or intestinal diseases. Rapid and accurate laboratory detection of infectious pathogens is important to either confirm or rule out an infectious process. A wide range of bacteria are associated with gastroenteritis. Here, we focus on the bacteria that are most commonly encountered in clinical practice and with a high clinical importance and/or frequency. In the Netherlands, the most common enteropathogens causing acute diarrhoea are *Campylobacter* species (mostly *C. jejuni*), less frequent enteropathogens are *Salmonella* spp, *Shigella* spp, and *Yersinia* spp. Also, a few outbreaks with the shiga-toxin producing *E. coli* (STEC) have been described in the Netherlands. At a minimum, routine diagnostic setup should be designed for the detection of *Campylobacter* and *Salmonella*. Diagnostics for *Shigella* is recommended upon symptoms after a stay outside Europe, diagnostics for *Yersinia is indicated* by persistent diarrhoea or stomach ache and detection of STEC is recommended only when an infection is suspected. In case of hospitalization and/or antibiotic use in the past three months, detection of *Clostridium difficile* can be considered.

At this time, the traditional algorithm for the identification of enteric pathogens is stool culture. Bacterial stool culture is limited by a low yield, because bacteria are difficult to isolate in the presence of normal stool flora. In addition, several enteric pathogens require highly selective media and will not be optimally recovered by culture. Therefore, molecular diagnostic techniques, such as PCR, are increasingly being designed and used in laboratories for the detection of bacterial enteropathogens.

In the following section, the most important enteropathogens will be discussed, including the relevance of molecular diagnostics for detection:

***Campylobacter* spp.** *Campylobacter* is one of the leading causes of bacterial diarrhoea. *Campylobacter* inhabits the gut of food animals as a commensal and meat can be contaminated during slaughter. *Campylobacter jejuni* and *Campylobacter coli* are established causes of gastroenteritis and clinically indistinguishable from each other. There is also compelling evidence from different European studies that emerging *Campylobacter* species are likely to contribute to the aetiology of gastroenteritis. A Dutch study showed that that 71.4% of 493 gastroenteritis cases were Campylobacter positive by PCR, with 4.1% positive for *C. jejuni*. Sequencing of the samples allowed identification of emerging Campylobacter species, including *C. concisus* (4.1%), *C. concisus or C. curvus* (0.8%), *C. ureolyticus* (0.6%), *C. gracilis* (0.6%), *C. showae* or *C. rectus* (0.4%), *C. upsaliensis* (0.4%), *C. hominis* (0.2%), and *C. sputorum* (0.2%). Many diagnostic laboratories fail to detect emerging Campylobacter species owing to the lack of specialized detection techniques. This is due to the fact that these fastidious organisms require micro-aerobic growth on special solid growth media, which makes detection by culture difficult. Culture-independent tests are increasingly being used for the detection of *Campylobacter* species, including sequencing techniques or genus- or species-specific PCR amplification, like 16S rRNA PCR. However due to the sequence similarity among *Campylobacter* species, the 16S rRNA cannot be used to differentiate between *C. jejuni* and *C. coli*. The larger 23S rRNA gene and the internal transcribed spacer (ITS) region have been used to make the distinction. It is shown that the ITS region is the most discriminatory region for species and strain differentiation for the *Campylobacter* genus.

***Salmonella* spp.** The majority of *Salmonella*-related disease occurs due to the ingestion of contaminated food or water. *Salmonella* can be divided into serotypes according to structures on the surface of the bacteria. *Salmonella* possesses a complicated taxonomy and the most important serotype causing human disease is *Salmonella enterica* (with over 2500 serotypes). Traditional *Salmonella* detection methods are based on stool cultures using selective media and characterization of suspicious colonies by biochemical and serological tests. These methods are generally time-consuming. Therefore, PCR based methods were designed for a more rapid identification of *Salmonella species* from stool. In addition, molecular typing methods (like PFGE, MLVA) can be used for subtyping of Salmonella, in order to identify and trace outbreaks. Serotyping of Salmonella is very laborious and expensive and is therefore usually better suitable for reference centres.

***Shigella* spp.** Transmission by humans is common for *Shigella*-related disease, because of a low infectious dose, only 10–100 organisms are needed to develop disease. The four species of *Shigella* are *Shigella dysenteriae*, *S. flexneri*, *S.boydii*, and *S. sonnei*. The most important risk factor for infection with *Shigella* is a stay outside Europe. Traditionally, *Shigella* detection is based on conventional stool cultures. However, the detection of *Shigella* by stool culture is difficult when compared to for example the detection of *Salmonella* by stool culture. For example, half of the culture-negative stool samples were positive for *Shigella* by PCR. Recently, laboratories started to combine conventional culture with molecular screening methods for the detection of *Shigella*. The PCR is used to screen for the presence of *Shigella* species DNA in the faeces prior to starting with the culture. However, as the *Shigella* target for PCR is difficult to distinguish from entero-invasive *E. coli*'s (EIEC), additional culture is still necessary. Also, PCR detection of *Shigella* must be considered in addition to stool culture, when there is a delay between taking of the sample and arrival of the sample at the laboratory, as the growth of *Shigella* can be diminished.

***Yersinia* spp.** Yersiniosis is most often caused by infection with *Yersinia enterocolitica, as* infection with *Yersinia pseudotuberculosis* is not frequently found. The pathogenic strains of *Y. enterocolitica* are identified by the serotype and biotype determination. Of the approximately 60 serogroups, only 11 serotypes are frequently associated with human disease (with serogroups O:3, O:8, O:9, and O:5,27 most dominant). Of the 6 different biotypes, five (biotype 1B, 2, 3, 4 and 5) are considered to be pathogenic.. Several different culture methods have been described for *Y. enterocolitica*, but no single method can detect all different serotypes. Furthermore, it has been shown that molecular diagnostic methods, like PCR, indicate a higher prevalence *of Y. enterocolitica.*

Escherichia coli. *E. coli* has long been considered as part of the human intestinal commensal flora, however some strains are proven to be pathogenic. The pathogenic strains are divided into different pathotypes. These pathotypes, also referred as diarrheagenic *E. coli* are enteropathogenic *E. coli (*EPEC), Shiga toxin-producing *E. coli* (STEC), enteroinvasive *E. coli* (EIEC), enterotoxigenic *E. coli* (ETEC), enteroaggregative *E. coli* (EAEC), and diffusely adherent *E. coli* (DAEC). Although *E. coli* is easy to identify at species level, recognition of the different pathotypes requires special laboratory techniques. The only pathotype identified mostly in routine diagnostics is the STEC (also called an enterohemorrhagic E coli (EHEC) or verocytotoxic producing *E. coli* (VTEC), which produces Shiga toxins [Shiga toxin 1 (Stx1) and/or Shiga toxin 2 (Stx2)]. The large outbreak in Europe 2011 was caused by the STEC O104:H4, frequently referred as EHEC. Around 5% till 10% of patients diagnosed with STEC infection develop haemolytic uremic syndrome (HUS), a life-threatening complication. Timely and accurate diagnosis of a STEC infection is important, because right treatment might improve patient outcome. This as antibiotic therapy in STEC infections might increase the risk to develop HUS.

Many STEC infections are missed due to current testing strategies. The use of conventional culture agars (like Sorbitol MacConkey SMAC-agar) only for diagnosing STEC infections results also in missed infections, as non-O157 infections are missed. The increased use of enzyme immunoassay (EIA) or PCR methods to detect the Stx-genes improved STEC diagnostics. Although non-culture tests are useful for diagnosis, a pure culture is still needed for serotyping and molecular characterization (e.g. pulsed-field gel electrophoresis patterns), which are essential for detecting, investigating, and controlling STEC outbreaks.

Clostridium difficile. C. difficile diarrhoea is a hospital-associated diarrhoea. *C. difficile* can be divided in toxogenic stains, which produce toxins, and non-toxogenic strains. The toxogenic strains are associated with diarrhoea. Exposure to antibiotics and health care facilities are the main risk factors for *C. difficile* infection. Therefore C. difficile diagnostic is recommended for patient with diarrhoea and recent (<3 months) antibiotic use or hospital stay. Also, *C. difficile* testing should not be performed on infants (i.e. patients <1 year of age) due to their high colonization rate. Detection of *C. difficile* by culture is possible, but should be combined with detection of the toxins. Detection of the toxins can be performed by using EIA or PCR.

1.3.2 Other Bacterial Pathogens

Most authoritative documents list *Aeromonas* spp. as accepted enteropathogens; however there are currently no reports of outbreaks of gastroenteritis with this bacteria. For diarrheal disease with *Plesiomonas shigelloides*, a definitive causal relation has yet to be established. Furthermore, over 60 *Vibrio* (-like) species have been described, but only a few are associated with bacterial gastroenteritis, with the two major species being *Vibrio cholera* and *Vibrio parahaemolyticus. V. cholera* can cause outbreaks at a pandemic level, especially in Africa and the southern regions of Asia. *V. parahaemolyticus* is responsible for food associated gastroenteritis outbreaks worldwide. Many laboratories include detection of *Aeromonas* spp., *Plesiomonas* spp., and *Vibrio* spp. into their routine diagnostics, whereas others add this only on request.

Bacillus cereus food poisoning can be caused by the ingestion of spores, which can germinate in food if not correctly cooled or heated. Another common bacterial foodborne disease caused by spores is *Clostridium perfringens*. Food poisoning *by Staphylococcus aureus* intoxication is caused by the ingestion of an enterotoxin. Food poisoning caused by *B. cereus*, *C. perfringens*, and *S. aureus* are infrequently diagnosed by clinical microbiology laboratories. This is likely due to the fact that few patients seek medical intervention for their symptoms, which are short-lived and detection is not always part of the testing algorithm. In cases of outbreak investigations, material could be sent to public health officials. *Listeria monocytogenes* can also cause intestinal infections. The major risk factor for developing gastroenteritis is the consumption of food items that are heavily contaminated (10^7–10^9 CFU/g or ml) with *L. monocytogenes*. Laboratory diagnosis is extremely difficult at this time, as there is currently no standard protocol for the detection of *Listeria* from human stools. Therefore *L. monocytogenes* is often not part of routine diagnostics in clinical microbiology laboratories. This testing is better suited to reference laboratories

1.3.3 Molecular Diagnostics Algorithms for the Identification of Enteric Pathogens

Molecular diagnostic tests have consistently demonstrated excellent sensitivity when compared to bacterial culture. There is a great promise for molecular diagnostics (mostly real-time PCR) for gastrointestinal pathogens when enhanced sensitivity is desirable and diagnostic yield of the conventional stool culture is poor. In addition, the use of molecular test can provide more specificity compared to conventional stool culture and it is a rapid technique, results are usually known within 6–8 h.

A very critical step is the extraction of intact nucleic acid from the sample in a molecular-based test. DNA extraction directly from stool is challenging. Stool is a complex mixture, rich in background nucleic acid and factors that could inhibit amplification. These substances can interfere with cell lysis or inactivate the DNA polymerase, and DNA extraction procedures are usually necessary to remove them. For example, the detection of a known amount of target DNA is often several logs reduced when this is placed in stool. Often, the stool samples have been pre-treated (for instance through cooling, heating, bead beating before extraction to limit the amount of inhibiting factors). There is some evidence that magnetic bead-based extraction systems yield less inhibitors than column-based systems. However, the inclusion of an extrinsic control before extraction is necessary to identify inhibition and possible false-negative results.

A disadvantage of singleplex real-time PCR is that it is most often targeted at one specific organism. Other potential enteropathogens that have not been included in the PCR are easily missed as individual testing directed on clinical symptoms is highly difficult as symptoms are not indicative of a specific bacterium. Because of the wide variety of potential bacterial (but also viral or parasitic) pathogens that can cause diarrhoea, there is a need to efficiently amplify multiple targets simultaneously. A lot of laboratories started to develop in-house multiplex PCRs for the detection of gastrointestinal bacteria. Multiplex PCR allows for the detection of multiple targets in a single reaction. Discrimination can be achieved by using different fluorescent reporter molecules, such as sequence-specific probes in real-time PCR. The advantage of testing one sample for multiple enteropathogens using a single test is a reduced turnaround time, the ability to identify co-infections and the high negative predictive value of a multiplex assay to avoid unneeded infection control measures or the unnecessary use of antibiotics. Although multiplex PCR is found to be a sensitive and quantitative detection method, its multiplexing capacity has been limited by the availability of fluorescent dyes and platforms. Also, sensitivity can be diminished by competition for available nucleic acid substrates or the reaction of substrates between the targets. Another option for the detection of multiple bacterial species is by using multiple singleplex PCRs. Here, the PCR is not limited by the available variety of fluorescent dyes or reaction substrates and might have improved sensitivity. However, performing a lot of single PCR reactions is extremely costly and time consuming compared to multiplexing. Recently, commercial molecular multiplex test for infectious diseases have gained popularity. There are two main categories of multiplex test. A multiplex syndromic (bacteria, virus and parasites) panel usually includes most common pathogens responsible for gastroenteritis. The other class included only a class of pathogens in a test (for example only bacteria). These commercial multiplex assays differ in many aspects, including the number of pathogens they detects, the throughput, the time to result and the detection format of the assay. An example is the Luminex xTAG GPP assay, the first large multiplex syndromic panel that received FDA approval. The xTAG GPP detects 14 of the most common pathogens (8 bacterial, 3 viral and 3 parasitic targets) from a single stool sample. The workflow of the assay starts with a sample pre-treatment with bead beating to break the cell wall of some pathogens, followed by nucleic acid extraction and a single multiplex RT-PCR. This is followed by a hybridisation reaction where specific tags on the amplified PCR products hybridize with their complement tag on the used beads. Detection of the beads is possible by signal detection of the fluorescent reporter that binds only to the amplified product in a flow-cytometry based way.

Although the potential for increased diagnostic yield is substantial, the clinical significance of isolated PCR findings can become less clear. Sometimes, the results of these tests are of unclear clinical significance, raising the question whether the detected target must be seen as a pathogen or not. One of the issues is that molecular test can give a positive results upon detection of DNA from non-viable pathogens or by non-pathogenic levels of a specific pathogen. Quantitative detection may prove useful to infer clinical significance. The hypothesis for the quantitative detection of DNA is that pathogens present at high concentration are more likely to be associated with disease. Quantitative detection can be performed by using real-time PCR. For quantification, the use of a standard curve of a known amount of target and extrinsic controls to check for the amplification efficiency are necessary. However, quantification of stool encounters problems not present in other matrixes, like plasma. This is because stool contains a lot of PCR inhibitors which could inhibit the controls needed for quantification. Or it is possible that the target is inhibited while the control is not, which would introduce incorrect quantification results. In short, quantitation of bacterial pathogens still needs more clinical evaluation since the relationship between pathogen burden and patients symptoms is still not evident for many pathogens.

It is clear that the use of molecular assays for bacterial pathogen diagnostics can improve sensitivity, and time to result with regard to conventional culture. However, that it will replace conventional cultures is still controversial. The main reason is that an isolate is still necessary for determination of phenotypic antimicrobial resistance or to be used for typing purposes. In some situations, a combination of conventional and molecular methods is used and proven to be effective. In these scenarios, PCR is used as a high-sensitivity screening test to select those samples that need further conventional testing. For example, the molecular assays rule all negative cases out for further testing, the conventional methods are used to confirm positive cases and for further determination tests. However, molecular positive but conventional negative samples need further confirmation tests. Ultimately, due to the reducing costs of whole genome sequencing and automation of data analyses, sequencing have become more and more common in labs for the characterization of isolates and is becoming routine practice.

1.4 Bacterial Typing

The final section of this chapter will discuss bacterial typing. Typing is the identification of different types of organisms within a species. Typing of microorganisms can be performed with a wide array of methods and depending on the typing goal, different methods can be employed. Within the hospital environment, typing can be employed to determine whether person-to-person transmission of a pathogen has occurred. In addition, it can also be used to make the distinction between treatment failure or re-infection. Recurrence of an infection in a patient can happen by two means: relapse of the primary infection that the treatment has failed to eradicate or re-infection with an unrelated exogenous strain. Typing of the microorganism in these circumstances can give more information. However, typing data by itself has little value, as it needs to be combined with clinical patient data, epidemiological information or demographic data to become meaningful.

On a national, international or even global scale, typing can be used in similar ways: to identify the source of an infection/outbreak or to identify transmission networks, but it can also be used to analyze the spread and expansion of specific variants of certain pathogens in surveillance studies. This can mean the spread of hyper virulent clones, but also the assessment of the effects of intervention methods, such as for example vaccinations. These latter studies tend to take place over a longer period of time and the pathogens will need to be typed both on strain and population level.

Microbial typing methods allow the characterization of bacteria on a strain level. Traditional typing methods were based on phenotypic characteristics of the bacteria. These methods can differentiate between organisms that show clear phenotypic differences, but infections can be caused by a small subpopulation of strains that show little phenotypic differences. In addition, some fastidious organisms cannot be cultured, making phenotypic characterization impossible. DNA based typing methods do not have (all of) these limitations and are widely employed, sometimes in combination with phenotypic methods.

Consideration should be given to the limitations in both the applicability of typing techniques and the typing results. For example, plasmid typing (i.e. categorisation of bacteria based on their plasmid content) is only possible for organisms that actually possess plasmids. Obviously, the discriminatory power should be sufficient to reveal differences between strains belonging to different species as well as between strains belonging to the same species.

Because a wide array of typing techniques is available, the selection of the appropriate technique to use can be made based upon several performance characteristics, in combination with the goal of typing. The performance of a typing technique is usually described by several characteristics: the typeability, the discriminatory power, the reproducibility, the flexibility, the portability and the ease of interpretation. The typeability indicates how well a method is able to type all the isolates studied. The discriminatory power indicates the resolution of the method to discriminate between unrelated isolates. The method should be able to reproduce the same results regardless of when it was tested. Preferably the method should be flexible, so it can type many different species using the same methodology. A method is called portable when it can be easily transferred between different systems or laboratories. Finally, the typing results should be easy to interpret. So, in general the ideal typing method should have optimal typeability, a high degree of reproducibility, be flexible, portable and have unprecedented discriminatory power and preferable for low costs and high throughput. But other characteristics can also be important, for example, when typing is used for international surveillance purposes, the data must be portable and preferably use internationally standardized nomenclature.

As the perfect test for all bacterial species at very low costs currently is not available, it is important to understand the advantages and limitations of the different methods when selecting the appropriate approach.

1.4.1 Typing Methods

Among the genotypic typing methods a rough distinction can be made between band-based methods (e.g. MLVA, AFLP, PFGE) and sequence based methods (e.g. Spa, MLST, whole genome sequencing). In general, the band based methods were considered faster (except for PFGE), as PCR is the underlying technique and the procedure can be finished within a day. This makes these techniques ideal when a rapid result is needed, for example in an outbreak investigation in a hospital. However, the reproducibility of some of these techniques tends to pose a problem and the portability is not as good. Recently, the introduction of automated band separation on capillary sequencers has increased the reproducibility and portability remarkably.

In the sequence based methods, the results are straightforward, increasing the portability and reproducibility. Sequencing used to be more time-consuming and costly then band-based methods, but the advances in whole genome sequencing methods and equipment has remarkably decreased the costs. The bioinformatics and data-analysis now seem to be the main limiting factor, but rapid advancements in automated data-analyses are currently being made to address this.

1.4.2 Pulsed-Field Gel Electrophoresis (PFGE)

In PFGE, genomic DNA is isolated and cleaved with a restriction enzyme that cuts infrequently, resulting in large fragments of DNA. These fragments are separated on an agarose gel by pulsed field electrophoresis, where the orientation of the electric field changes periodically. In PFGE a large portion of the genome is assessed and large genomic events, such as recombination, insertion or deletion of large mobile elements will result in a change in the PFGE pattern. In addition, large plasmids can also be seen on the gel. Pulsed field gel electrophoresis has been considered the 'gold standard' of molecular typing methods for a wide spectrum of clinically relevant pathogens. PFGE has been the primary approach for outbreak characterization and the analysis of transmission events, but dependent on the causative organism, other higher throughput methods, like MLST, MLVA and next gen sequencing are replacing this method more and more. However, as it is a relatively inexpensive, but time-consuming method, with high discriminatory power, good epidemiological concordance and excellent reproducibility, it will remain the method of choice in low-resource settings until other methods become less expensive. In addition, standardized protocols (through Pulsenet and Harmony) and quality control panels are available for some bacterial species, increasing the portability and quality control.

1.4.3 Amplified Fragment Length Polymorphism Typing (AFLP)

In the AFLP method, genomic DNA fragments resulting from restriction with (one or) two enzymes are specifically ligated to double stranded adaptors. These restriction fragments with ligated adaptors are then selectively amplified by PCR, using primers complementary to the adaptor sequence, the restriction site sequence and a specific number of extra nucleotides. The PCR is performed with highly stringent conditions at the beginning, ensuring efficient primer binding to the complementary nucleotide sequence of the template. In this way, large numbers of restriction fragments are amplified in a single reaction. The reaction is usually performed with one fluorescently labeled primer to facilitate separation on an automated DNA sequencer with subsequent computer-assisted comparison. The greatest advantages of this method are the good discriminatory power and the flexibility: with only a limited set of reagents, using the same procedure, a wide variety of organisms can be typed. Developments in this technique have decreased the turn-around time to 24 h, making this a suitable method when rapid results are needed (Fig. 1.1).

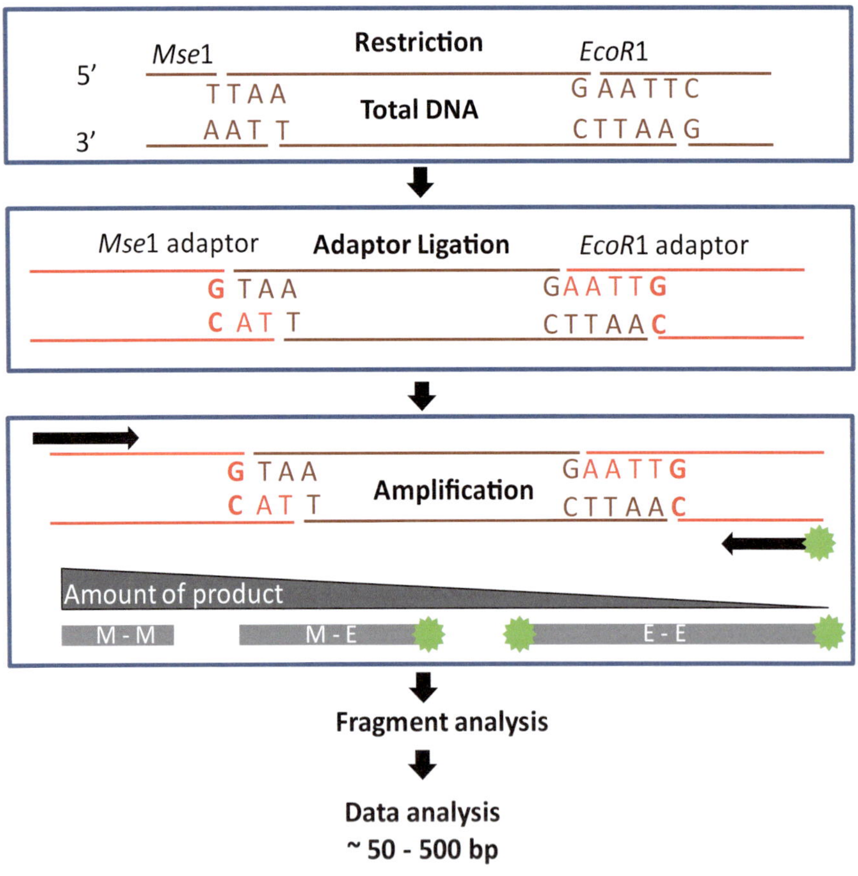

Fig. 1.1 Amplification Fragment Length Polymorphism (AFLP)

1.4.4 Variable Number of Tandem Repeat (VNTR) Typing

In bacterial genomes, many regions with repetitive nucleotide repeats can be found, both in coding and non-coding regions. These repeats can vary from a few bases to more the 100 base pairs in length and when the repeats are directly adjacent and their number at the same locus varies between isolates, the region is called a variable number of tandem repeat locus. These repeats are prone to mutation due to slipped strand mispairing during replication. In VNTR typing, differences in the number of repeats are assessed to distinguish isolates using PCR based techniques and separation on an automated sequencer. To increase the discriminatory power, frequently multiple loci are used, in a multiple locus VNTR analysis (MLVA). This method is used for a wide array of organisms. Pulsenet harmonization protocols are available for *E.coli* O157 and *S. enterica* serotypes Enteritidis and Typhimurium. Other (curated) databases exist for MRSA, *Haemophilus influenza* and many other organisms. Typing by MLVA is fast, unambiguous and discriminatory. The main limitation is that the method is not universal, as primers for each pathogenic species need to be developed. In addition, harmonization of the allele amplicons is very important for the reproducibility of the data.

1.4.5 Random Amplification of Polymorphic DNA (RAPD)

RAPD-PCR is based upon the use of short random primers that hybridize to several unspecified DNA sequences. The resulting amplicons can be analyzed on agarose gel electrophoresis or on a DNA sequencer to generate a bacterial fingerprint, which can be compared. The number and positions of the primer binding sites is specific for a particular bacterial strain. This method is rapid, simple and inexpensive, but the intra- and inter-laboratory reproducibility is low. This method is also less discriminatory than PFGE. Variants of RAPD-PCR, (e.g. arbitrarily primed PCR (AP-PCR)) are based on the same principle, but have differences in primer length and amplification protocol.

1.4.6 Other Amplification Based Methods

Other PCR-based techniques used for typing address certain (repetitive) regions in the genome which can be used for typing. Some examples: enterobacterial repetitive intergenic consensus (ERIC)-PCR, repetitive extragenic palindromic (rep)-PCR and the PCR melting profile (PCR-MP) technique.

1.4.7 Single Locus Sequence Typing

Sequence based typing methods can be used to determine the relation between bacterial isolates based on the sequence variability in one or more loci. Single locus sequence typing (SLST) is used in organisms where a certain gene shows a high rate of variability, usually because the gene product is surface exposed. In SLST, the locus of interest is amplified by PCR and subsequently sequenced. Sequences can then be compared to online database to determine a type. For example, typing of group A *Streptococcus* can be performed based on the sequence of the *emm* gene, which encodes the surface M-protein, a major virulence and immunological determinant of this pathogen. For *Campylobacter*, the short variable region of the *flaB* gene, encoding the minor flagellin subunit FlaB, can be used to study *Campylobacter* epidemiology. *Neisseria gonorrhoeae* can be typed by full—or extended-length sequencing of the *porB* gene, but multi-locus sequence typing is more common for this organism. As highlighted above, typing of *M genitalium* has been important in demonstrating transmission networks and is usually performed by sequence typing of the *mgpB* gene. In these organisms, SLST can be used for epidemiological studies, but to fully discriminate closely related organisms, for example in outbreak situations, it usually has to be combined with another typing method as the single locus sequencing does not give enough discriminatory power in closely related isolates. In general, this method can be performed easily and the results are easy to interpret and portable, however, the resolution and costs will not be acceptable in certain settings.

1.4.8 Spa Typing

A special form of SLST is the *S. aureus* protein A gene (Spa) typing, in which the polymorphic X region of this gene is amplified and sequenced. This methods relies on the identification of the short repeats in this region, which vary in number of repeats and point mutations within the repeats. Each repeat is assigned a code and the spa type is deduced from the order of the specific coded repeats. The discriminatory power of spa typing is lower than PFGE, but it is very fast, cost effective, easy to use and has standardized international nomenclature. The major drawback is the fact that it is a single locus technique and can misclassify isolates. This is illustrated by the fact that recently, in the Netherlands, the national institute of public health (RIVM) decided that MLVA typing is to replace the spa typing for the national MRSA surveillance. According to the data generated over several years of performing simultaneous spa and MLVA typing, it was shown that MLVA is superior to spa typing and that MLVA will suffice to characterize MRSA isolates for surveillance.

1.4.9 Multilocus Sequence Typing

In multilocus sequence typing (MLST) the genetic relatedness is assessed using multiple genes. Here, multiple loci on the genome are amplified by PCR and subsequently sequenced. As multiple loci are analyzed simultaneously, the discriminatory power is higher than SLST. Depending on the varying degrees of genetic drift, which can be gene type and organism specific, various types of genes can be chosen to be part of the MLST scheme. Schemes using housekeeping genes are most common, as these do vary to some extent, but are stable enough to be amplified by PCR. However, the limited extent of variation also limits the discriminatory power. Typing systems using polymorphic genes (usually genes under selective pressure, such as virulence genes) have also been developed to increase the discriminatory power. For example, for *C. trachomatis*, several multi locus sequencing typing systems that vary in both the number and type of loci used (housekeeping versus polymorphic genes) have been developed due to the need for genotyping methods with high resolution. For *Neisseria gonorrhoeae*, a two-locus MAST (multi antigen sequence typing) scheme using two very variable genes has been developed, which gives better discriminatory power than the SLST of the *porB* gene. Schemes employing combinations of housekeeping and polymorphics genes have been developed as well, as for *Salmonella enterica*, an MLST has been developed that combines two housekeeping genes with two variable flagellin genes. In MLST, the unique sequences of each loci (alleles) are usually assigned a number and the combination of alleles or allelic profile results in a sequence type which can then be determined using international online databases. The great advantage is that this leads to unambiguous typing results that are highly reproducible. The greatest disadvantages are that the method is labor intensive and time-consuming, combined with the (sometimes high) cost of sequencing. In addition, for some pathogens this method does not have enough discriminatory power for routine use in outbreak settings.

1.4.10 Optical Mapping

In optical mapping an ordered genome-wide restriction map of selected individual DNA molecules is created. HMW DNA is isolated and stretched on a microfluid device. The DNA is digested with a restriction enzyme, and as the DNA remains attached to the surface the molecular fragments remain in the original genomic order. Special software can create a consensus genomic optical map by overlapping single molecule maps. The resulting maps have a high resolution, facilitating assessment of genetic relationships among bacteria. However, specialized equipment is needed and mapping of a few organisms can take up to 2 days.

1.4.11 Whole Genome Sequencing (WGS)

Whole genome sequencing, using next gen sequencing techniques, has made strong advances in recent years. A whole genome of a bacterium can be sequenced within one hour (excluding preparation and data-analysis time). To construct the complete genome from the data generated, all the reads need to be assembled based on overlapping regions (de novo assembly) or compared with previously sequenced reference genomes. The high amount of data generated makes this a very powerful tool for epidemiological investigations. Currently, the main drawback is the time and computational power necessary for data analysis. The availability of web-accessible bioinformatics platforms for data processing and analysis would facilitate the use in clinical settings. Furthermore, comparison between conventional typing methods, such as PFGE and MLST, and whole genome data would be ideal. This will become more and more routine over the next few years, as automated extraction of typing information for bacterial pathogens from whole genome sequence data is currently being developed.

1.4.12 Typing for Action

Typing results, in combination with clinical, epidemiological or other data, can be used to rapidly and effectively direct medical and public health action. Many techniques are available, each with their advantages and limitations, depending on the purpose of the study. The choice of methods relies on the intrinsic properties of the method, but also on the cost, ease of interpretation, as well as on the cost of setting up and maintaining a method in the lab. The skill level and resources can also pose limiting factors on the choice of method, as can the need for international standardization of a technique.

WGS data are increasingly used to characterize and type bacterial pathogens. These data can provide detailed information on the genotypes and likely on the phenotypes of microorganisms. This can help to establish relationships of disease outbreak samples with more precision. However, although whole genome sequencing does solve problems like typeability, the acquired data does not necessarily answer all typing questions in the most effective way. In some circumstances, for example a hospital outbreak, the resolution of next gen sequencing might not be necessary when simpler and faster methods can demonstrate transmission just as well. There, speed might take precedence over discriminatory power. In other settings, the increased resolution of whole genome sequencing, which gives the opportunity to look at resistance genes and possible mutations in certain virulence genes, could be more appropriate. Based upon the exponential development of the next generation sequencing methods in combination with the cost decrease for full-genome sequencing, it could be speculated that this might become the sole typing scheme adopted by researchers. However, whether this would be the method of choice in a hospital setting, where the ability to rapidly decide where and when control measures need to be implemented is more important than the highest typing resolution, remains to be seen. Lastly, this overview of techniques is in no way complete and many other useful typing methods exist as well.

1.5 Conclusion

In this chapter a non-comprehensive overview of molecular diagnostics and typing methods of clinically relevant bacterial pathogens has been described. The two examples used here were aimed to highlight the different available methods for the diagnosis of bacterial pathogens. Naturally the field of molecular diagnostics of bacterial pathogens includes many other important pathogens. Molecular techniques for the detection of pathogens causing respiratory tract infections are starting to replace or being used in addition to culture based methods. Several commercial molecular tests for respiratory tract infections are available. Assays for the rapid molecular testing of bloodstream infections have also been developed. In addition, the determination of the antimicrobial susceptibility of clinical isolates, especially with the increase in antimicrobial resistance observed worldwide, is crucial for the optimal antimicrobial therapy of infected patients. Molecular techniques for the detection of resistance may offer advantages over phenotypic assays and are currently widely in use for certain pathogens, such as the detection of the methicillin resistance-encoding *mecA* gene in *staphylococci* and rifampin resistance in *Mycobacterium tuberculosis*.

The field of molecular diagnostics of bacterial pathogens is constantly moving and new techniques are being developed and implemented in a continuous search for assays with an even higher analytical sensitivity and/or specificity. In addition, the increasing automation of assays is facilitating high-throughput analysis and reduction of errors or contamination during manual handling. With these developments, the importance of molecular diagnostics of bacterial pathogens for clinical microbiology laboratories will likely only further increase in the future.

Recommended Literature

Centers for Disease Control and Prevention. Recommendations for the laboratory-based detection of Chlamydia trachomatis and Neisseria gonorrhoeae. MMWR Recomm Rep. 2014 Mar 14;63 (RR-02):1–19.

Espy MJ, Uhl JR, Sloan LM, Buckwalter SP, Jones MF, Vetter EA, Yao JD, Wengenack NL, Rosenblatt JE, Cockerill FR 3rd, Smith TF. Real-time PCR in clinical microbiology: applications for routine laboratory testing. Clin Microbiol Rev. 2006;19(1):165–256.

Fournier PE, Dubourg G, Raoult D. Clinical detection and characterization of bacterial pathogens in the genomics era. Genome Med. 2014;6(11):114.

Jensen JS. Protocol for the detection of Mycoplasma genitalium by PCR from clinical specimens and subsequent detection of macrolide resistance-mediating mutations in region V of the 23S rRNA gene. Methods Mol Biol. 2012;903:129–39.

Trembizki E, Costa AM, Tabrizi SN, Whiley DM, Twin J. Opportunities and pitfalls of molecular testing for detecting sexually transmitted pathogens. Pathology. 2015;47(3):219–26.

Zhang H, Morrison S, Tang YW. Multiplex polymerase chain reaction tests for detection of pathogens associated with gastroenteritis. Clin Lab Med. 2015;35(2):461–86.

Chapter 2
Molecular Diagnostics of Parasitic Infections

Jaco J. Verweij

Abstract DNA-based methods have been available for many years in most clinical microbiology laboratories. However, until recently, these tools were not routinely exploited for the diagnosis of parasitic infections. Laboratories were reluctant to implement PCR, not knowing how to incorporate such an approach in the algorithm of tools available for the most accurate diagnosis of a large variety of parasites. This was especially noticeable in the diagnosis of intestinal parasitic infections, where the diversity of parasites that one can expect in most settings, is much smaller than the parasitological textbooks suggest. Therefore, the classical algorithm, which is based on population, patient groups, use of immuno-suppressive drugs, travel history, etc., is also applicable to the decision on whether or not to perform additional techniques when a multiplex PCR panel is used as a first line diagnostic.

Keywords Protozoa · Helminths · Malaria · *Giardia lamblia* · Intestinal parasites

2.1 Introduction

Traditionally, laboratory diagnosis of parasitic infections is based on the examination of microscopical slides and, as a consequence, highly depended on the expertise and skills of the microscopist. Although various staining and concentration techniques have been introduced to enhance the specificity and sensitivity, the diagnostic performance is known to be low and time-consuming. Very soon after the first publication on in vitro DNA amplification in a polymerase chain reaction (PCR), a breakthrough was predicted in molecular parasitology and in the diagnosis of parasitic infections. Until that time, specific DNA probes were used in fundamental research, but diagnostic applications had been hampered by the limited sensitivity of these hybridization techniques. In 1995, a number of publications for

J.J. Verweij (✉)
Laboratory for Medical Microbiology and Immunology, Elisabeth-TweeSteden Hospital, Tilburg, The Netherlands
e-mail: j.verweij@etz.nl

E. van Pelt-Verkuil et al. (eds.), *Molecular Diagnostics*,
DOI 10.1007/978-981-10-4511-0_2

DNA-based methods for the detection and identification for a range of parasitic infections were reviewed by Judith Weiss. At that time, the use of PCR was still limited and most of the research was done on malaria, *Leishmania*, trypanosomes and *Toxoplasma*, which are all tissue parasites. Moreover, except for toxoplasmosis, experience with PCR performed on DNA that had been isolated directly from patient material was scarce.

In the last 10 years, many diagnostic laboratories have been provided with the facilities to perform molecular diagnostics. The introduction of real-time PCR reduced the problems related to the contamination of samples with PCR products and it became relatively easy to combine multiple DNA targets in one multiplex assay. In addition, the implementation of automated DNA isolation methods made it possible to use DNA-based detection techniques in a high-throughput format in diagnostic routine laboratories.

Molecular detection, differentiation and genotyping methods have been described for a large number of parasitic infections. These methods are now performed in both diagnostic and research settings. This chapter is a selection of the most clinically relevant parasitic infections in industrialized countries, with an emphasis on the use of molecular diagnostics in a routine setting.

2.2 DNA Isolation

Without a proper method for DNA isolation, DNA amplification techniques cannot be used. Especially when parasitic DNA has to be extracted from faeces, proper DNA extraction is extremely important. When setting up a PCR, it is essential to ensure that the target DNA will be released from the parasitic stage (e.g., cysts, eggs or spores) with the isolation method used and from the relevant clinical material. Moreover, amplification of DNA isolated from faeces is notoriously difficult due to the presence of factors that can inhibit the amplification reaction (see Vol. 1, Chaps. 5 and 6). Heating the faecal sample and the addition of absorbent substances such as polyvinyl polypyrolidone (PVPP) during the DNA isolation procedure or the addition of inhibition factor-binding substances, such as bovine serum albumin (BSA), in the PCR may be used to prevent inhibition of the amplification. However, it always remains important to use an internal inhibition control in each reaction. For example, Phocine Herpes Virus (PhHV) can be added during the DNA isolation procedure, after which a specific PhHV PCR, in multiplex with the PCR(s) for the target(s), can be performed. Nowadays, more and more DNA isolation kits and protocols for automated DNA isolation systems are commercially available. As highlighted above, with the introduction of a new DNA isolation protocol or when changing an existing protocol, it is vital to consider whether the DNA of the intended target(s) is released. This can be done by testing known positive samples. Both in our own experience and as published by others, preservation of faeces with formalin or saline acid formalin (SAF) has an enormously negative effect on the specific amplification of DNA. This effect is enhanced by the time that the sample has been stored in the fixative.

2.3 Intestinal Parasites

2.3.1 Protozoa

Entamoeba histolytica is the cause of amoebic colitis, amoebic dysentery and amoebic liver abscess and is responsible for an estimated 100,000 deaths per year. Until about 25 years ago, differences in clinical symptoms were explained by the existence of pathogenic and non-pathogenic strains of *E. histolytica*. Biochemical, immunological and genetic differences between these "strains", however, are so large that it is now generally accepted that they are in fact two different species, named *E. histolytica* and *Entamoeba dispar*. Because the potentially invasive *E. histolytica* is morphologically indistinguishable from the non-invasive *E. dispar*, microscopic examination of faecal samples only, is not suitable for a definitive diagnosis of *E. histolytica* infection and additional methods, such as the detection of specific antigens or specific DNA sequences are necessary. Although antigen detection is specific, its sensitivity is too low, when compared with PCR, for use in patient care.

Giardia Lamblia infections are very common, occur globally and are seen as one of the most important non-viral causes of diarrhoea in industrialized countries. Since the time of Anthony van Leeuwenhoek, the diagnosis has been based on microscopic investigation of (multiple) faecal samples using concentration techniques. The detection of *Giardia* antigens by ELISA has been accepted in recent years as a cost-saving alternative. PCR targeting the SSU rDNA has excellent sensitivity and specificity when compared to microscopic and antigen based detection strategies. *Giardia* can be divided into various genotypes (assemblages) on the basis of Restriction Fragment Length Polymorphism (RFLP) analysis and DNA sequence analysis of several (household) genes. However, the clinical significance of the various assemblages is still unclear and therefore, these genotyping methods are used in epidemiological and basic research settings only.

Cryptosporidium-associated diarrhoea is best known for the severe symptoms in AIDS patients during the HIV/AIDS epidemic in the 1980s and 1990s. Subsequently, improved diagnostic methods have been developed. Probably due to this awareness, *Cryptosporidium* is now also known as the cause of large water- and food-borne outbreaks of gastroenteritis in immunocompetent individuals. Modified acid-fast staining techniques are often used for the detection of *Cryptosporidium* oocysts in stool. However, the sensitivity and specificity appear to be low and proper identification is very dependent on the experience and skills of the microscopist. Although immunological detection methods are available, these are not as well accepted as in the diagnosis of *Giardia* infections due to low sensitivity and specificity. PCR based techniques have been proven to be a sensitive and specific alternative for the detection of *Cryptosporidium* in faecal samples. In a study including 950 patients from the Groningen region in the Netherlands, who visited their family doctor with gastrointestinal symptoms, *Cryptosporidium* was demonstrated using real-time PCR in 20% of children under five.

PCR based techniques have also been shown to have a high sensitivity for the detection of other coccidia-like infections, such as *Cystoisospora belli* (*Isospora belli*) and *Cyclospora cayetenensis*. Another example in which PCR is a worthwhile alternative in the diagnosis of opportunistic pathogens is microsporidiosis. Microscopic detection of the very small spores of different microsporidium species in faecal samples, even after additional staining techniques, has proven to be extremely difficult and time-consuming. Amplification of microsporidium-specific DNA has been shown to be a sensitive and specific tool in the diagnosis of microsporidium infections. Probably due to the increased attention on microsporidiosis during the AIDS epidemic and the improvement of diagnostic techniques, such infections are now being diagnosed more frequently in transplant patients, children, the elderly and in travellers.

Dientamoeba fragilis was first described in 1918 as a commensal living amoeba of the gastro-intestinal tract of humans. Subsequently, through antigen based and ultrastructure studies and through the analysis of the 16S ribosomal RNA, the organism was classified as a flagellate, though a flagellum is missing. Since its discovery, the pathogenicity of this organism has remained controversial. In recent years, several authors have published on the clinical significance of *D. fragilis* as a cause of gastrointestinal distress, unfortunately, a consensus on the pathogenicity is lacking. One of the issues preventing progress is that many *D. fragilis* infections remain without symptoms. Until recently, a cyst stage of *D. fragilis* was not recognized. Therefore, the fragile trophozoites were detected and identified by microscopic examination in fresh stool or by the use of fixatives and permanent staining techniques. The sensitivity of a single microscopic examination is not high, because the day-to-day variation of *D. fragilis* trophozoites in the faeces seems even more irregular than observed in other intestinal protozoan infections, such as *G. lamblia* and *E. histolytica*. Immunological techniques for the detection of *D. fragilis* are not available. Real-time PCR has been proven to be a highly sensitive and specific alternative to the diagnosis of *D. fragilis*, in which the influence of the day-to-day variation of the results is greatly reduced. *Dientamoeba* DNA can be detected successfully with qPCR, even after storage of the faecal samples for up to eight weeks at 4 °C, without loss of sensitivity. Genetic variations of the small subunit ribosomal RNA (SSU rRNA) gene and of the internal transcribed spacer 1 (ITS1) of *D. fragilis* have not been associated with the occurrence of clinical symptoms. Furthermore, in a recent multicentre case-control study in the Netherlands, *D. fragilis* was detected more often in controls as compared to diarrhoeal cases, which has led many laboratories to remove *D. fragilis* from their primary screening panel.

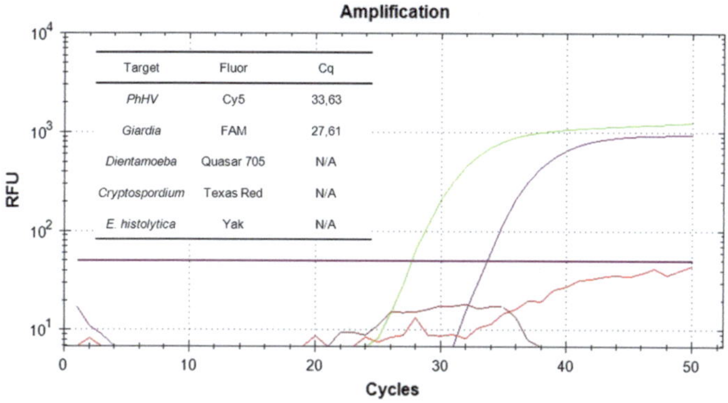

Fig. 2.1 Example of amplification curves of Giardia and the internal inhibition control in the same sample in a multiplex real-time PCR for the detection of *E. histoytica, D. fragilis, G. lamblia, Cryptosporidium* and PhHV DNA (N/A = Not detectable)

Blastocystis is the most common intestinal parasite in the faeces of both patients with gastrointestinal complaints and in asymptomatic individuals. It is thus, even more so than with *D. fragilis*, unclear whether it is necessary to detect this parasite. PCR-based techniques display a higher sensitivity when compared to microscopy and culture and provide the possibility of performing sub-typing by means of DNA sequence analysis. Factors that seem to be associated with the clinical significance of *Blastocystis* are the parasite load, the genotype or subtype, and the use of antimicrobial treatment (Fig. 2.1).

2.3.2 Helminths (Worms)

Strongyloides stercoralis is endemic in many tropical and subtropical regions. It is estimated that worldwide 50–200 million people are infected, with prevalence rates as high as 50%. *S. stercoralis* behaves differently from other intestinal nematode infections. Rhabditiform larvae are excreted in the faeces, but may already develop to infectious filarial larvae in the intestines. These larvae can then penetrate through the intestinal mucosa or the perianal skin, which results in a long-term cycle of auto-infection that can continue for many decades. Chronic strongyloidiasis may result in hyper infection, which is often seen in immunocompromised hosts, such as transplant patients, patients receiving chemotherapy and patients treated with corticosteroids.

The laboratory diagnosis of *S. stercoralis* is mainly based on serology and on the detection of *S. stercoralis* larvae by microscopic examination of faecal samples. Microscopy is very labour intensive, particularly in chronic infections. The number of larvae is very small and even after formalin-ether concentration (the Baermann method) or coproculture, sensitivity is low and multiple samples have to be examined in order to achieve the appropriate level of sensitivity. A *S. stercoralis*-specific qPCR was described, with very high specificity, using a large panel of DNA controls and faecal samples. The primer and probe set is based on the 18S ribosomal RNA gene sequence and turned out to be 10 to 100 times more sensitive in comparison with primers and probes designed on the cytochrome c oxidase subunit I gene and a *S. stercoralis*-specific sequence.

For various other soil-transmitted helminth (STH) infections, such as *Ascaris* and hookworm (*Necator americanus* and *Ancylostoma duodenale*), PCRs targeting of the Internal Transcribed Spacer 2 (ITS2) sequence have been developed, but these are still mainly used for epidemiological studies in endemic areas. The occurrence of these infections is very limited in the Netherlands (see also "Algorithm faecal examination") and infections are often very mild. In addition, these worms have a relatively short life span. It is important to realize that several of the isolation methods that have been published were proven to be unable to release DNA from *Trichuris trichiura* eggs. As a consequence, until recently, PCR for the detection of this parasite could not be integrated into a multiplex panel (Fig. 2.2).

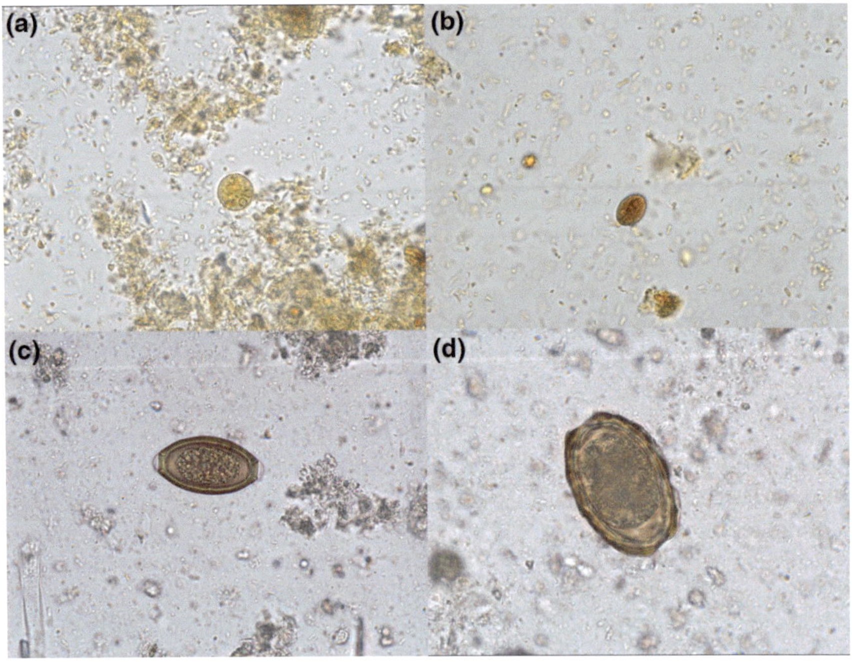

Fig. 2.2 Example of protozoan cysts: *Entamoeba histolytica/Entamoeba dispar* (**a**), *Giardia lamblia* (**b**) and worm eggs *Trichiura trichiura* (**c**), *Ascaris* (**d**). Photos are courtesy of Eric A.T. Brienen, Department of Parasitology, Leiden University Medical Center

2.3.3 Algorithm for Faecal Examination

Conventional methods for the detection of intestinal parasites have their limitations in specificity and sensitivity. This means that species-specific techniques or modifications for each individual organism are necessary to attain the most optimal diagnosis. Multiplex real-time PCR, however, provides a sensitive and specific detection for each of these parasites.

Which PCRs are used in the diagnosis of intestinal parasitic infections is very dependent on the patient population of which the laboratory receives faecal samples. However, the choices that are made do not differ substantially from those used in the decision on the use of additional concentration and staining methods to improve the sensitivity of the detection of specific parasites using microscopy. After all, a coproculture or Baermann for the concentration of *Strongyloides* larvae, for example, is used only when the travel or clinical history provides indications to do so. In the same way, one or more multiplex PCRs can be chosen. These can be supplemented with either PCR(s) for additional targets and/or additional species-specific microscopic methods.

Examples of various target combinations in specific groups of patients are shown in Table 2.1.

Table 2.1 Examples of targets in different patient populations

Adults	Children
Entamoeba histolytica[d]	*Dientamoeba fragilis*[e]
Giardia lamblia	*Giardia lamblia*
Cryptosporidium	*Cryptosporidium*
Travellers	**Immunocompromised patients**
Entamoeba histolytica[d]	*Strongyloides stercoralis*[a]
Giardia lamblia	*Cryptosporidium*
Cryptosporidium	*Cystoisospora belli*
Strongyloides stercoralis	*Enterocytozoon bieneusi*
Schistosoma[b]	*Encephalitozoon* spp.
Cyclospora cayetanensis[b]	
Cystoisospora belli[b]	
E. bieneusi[c]	

[a]Recommended before treatment with corticosteroids is started
[b]In travellers to high-risk areas
[c]Importance in travellers has been suggested
[d]Recommended before diagnoses of M. Crohn is made
[e]Clinical significance unclear

2.4 Blood and Tissue Parasites

2.4.1 Protozoa

Malaria

Although the number of malaria cases in non-endemic areas is decreasing, the diagnosis of malaria remains a challenge for diagnostic laboratories. The potential fatality of malaria tropica in a patient with fever returning from the tropics makes it necessary to put all other work aside to exclude or confirm an infection with *Plasmodium*. In most laboratories, microscopy and antigen detection are performed as the standard diagnostics, however, PCR has been accepted as a highly sensitive and specific method, in particular in the detection of sub-microscopic infections. Nowadays, PCR can be considered as the gold standard for the differentiation of *Plasmodium* species and for the detection of mixed infections. Until now, malaria PCR is not often performed as a first line diagnostics tool, it is merely being performed in a number of reference centres, for the confirmation of microscopic findings. This limited application is mainly the result of logistics; in most laboratories the microbiological and haematological analytical staff run 24-hour shifts and those specialized in molecular diagnostics usually do not. The 18S rRNA gene is used as the target in conventional and, more recently, real-time PCRs for the detection and differentiation of the four human malaria parasites *P. falciparum, P. vivax, P. ovale* and *P. malariae*. Recently, human infections with the "monkey malaria" parasite *P. knowlesi* have been described and real-time PCRs, targeting the 18S rRNA gene of this parasite, have been published. Because resistance to the various anti-malarial drugs is an ever increasing problem, various techniques for the detection of mutations in several genes and the number of copies of genes involved in the metabolism of the various anti-malarial drugs in the parasite are applied in the research setting and in epidemiological studies. Examples of such genes are the chloroquine resistance Plasmodium transporter gene (Pfcrt) and the Multi-Drug Resistant Plasmodium falciparum gene 1 (Pfmdr1).

Leishmania

The diagnosis of cutaneous and visceral leishmaniasis import infections has traditionally been based on the extensive and intensive microscopic examination of biopsy material or bone marrow by experienced laboratory staff. This is very labour intensive and can be extremely difficult. Nucleic acid amplification methods, such as (real-time) PCR, offer a very sensitive and specific diagnostic alternative. Moreover, these methods give the possibility of species differentiation which is important for the choice of treatment. A genus-specific qPCR based on the SSU rRNA gene is used for the detection of *Leishmania*-specific DNA. The higher sensitivity of molecular diagnostics makes lengthy microscopic examination unnecessary. In cases of suspected visceral leishmaniasis, PCR on blood instead of more invasive bone marrow examination is a worthwhile alternative, in addition to serological testing. Research in Sudan showed that with conventional PCR, *Leishmania* DNA is detectable in blood in 70% of the proven leishmaniasis patients. This sensitivity is even greater when qPCR is used in which only a very small target is amplified.

The taxonomy of *Leishmania* species is complex and the nomenclature of the different types of complexes is subject to change. This makes it sometimes difficult to compare differentiation techniques. Depending on the travel history, various PCR-RFLP and DNA-sequencing methods based on different targets, such as Internal transcribed spacer sequence (ITS), Heat Shock Protein 70 (HSP70), T2B4 repeat, 7SL and mini-exon DNA, are used for the differentiation of the different species and species complexes. DNA-sequencing of larger targets with higher

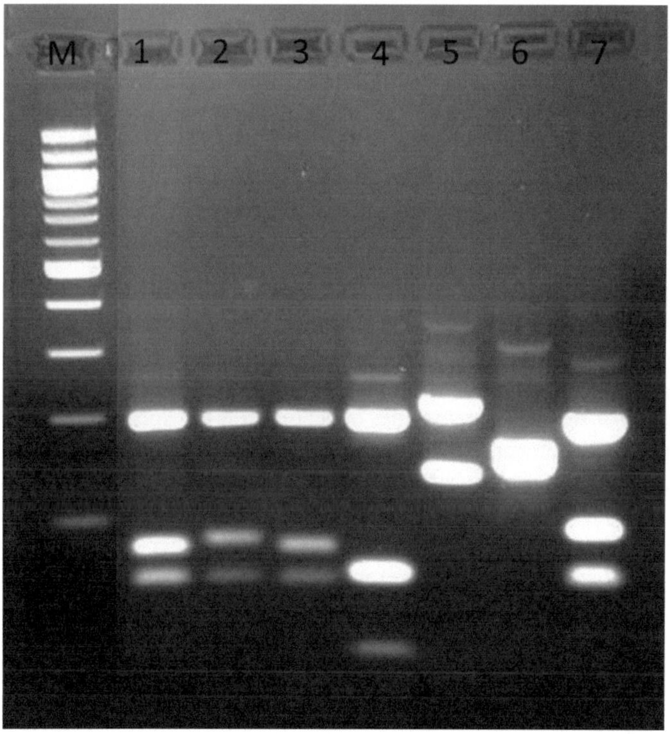

Fig. 2.3 Restriction Fragment Length Polymorphism (RFLP) analysis. Fragments Pattern of a *Leismania* Internal Transcribed Spacer PCR product after digestion with Hae III. 100 bp marker (M), no difference between *L. infantum* and *L. donovanni* (Lane 1–3), unique patterns for *L. tropica* (Lane 4), *L. major* (Lane 5). No difference within species belonging to the *L. Viannia* subgenus (Lane 6), *L. mexicana* (Lane 7)

genetic diversity, such as HSP70, results in higher resolution of the different species as compared to PCR-RFLP of smaller targets, e.g. ITS (Fig. 2.3).

Toxoplasma

The laboratory diagnosis of toxoplasmosis is primarily based on serology. The detection of specific IgG and IgM antibodies in otherwise healthy people will usually be sufficient for a toxoplasmosis diagnosis or for the determination of the immune status. When immunocompromised, in both ocular toxoplasmosis and congenital toxoplasmosis, this is often much more complicated and molecular diagnostics may complement the serological data. For example, in countries where serological screening for toxoplasmosis in pregnancy is common, *Toxoplasma* PCR is applied on amniotic fluid in cases of seroconversion. Real-time PCR on plasma also shows a simple, rapid and highly sensitive detection method for the early diagnosis of disseminated toxoplasmosis in bone marrow and organ transplant patients. In literature, a large number of DNA targets are described in *Toxoplasma*-specific PCRs, such as the B1, P30 and 18S ribosomal DNA gene and, more recently, a 529 bp-Toxoplasma-specific repeat. A comparison of the sensitivity and specificity of the different targets is difficult, because they are highly depended on the size of the PCR product, the primers, the detection probes and on the amplification and detection methods used.

2.4.2 Helminths

Schistosoma

The adult *Schistosoma* worms reside, depending on the species, in the blood vessels, around the intestines or around urinary bladder. Therefore, these are blood-borne parasites, strictly speaking. However, the eggs are excreted through the intestinal wall into the faeces by *Schistosoma mansoni* or through the bladder wall into the urine by *Schistosoma haematobium*. The number of eggs, especially for travellers, is mostly very low. Therefore, larger amounts of faeces and/or urine are concentrated to increase the sensitivity of the microscopic examination. Because of the mostly mild infections in the non-endemic setting, serological testing is more sensitive. In endemic areas, *Schistosoma*-specific PCR, targeting the internal transcribed 1 (ITS-1) sequence on DNA isolated from a small amount of urine or faeces, proved more sensitive than microscopic examination. Recently, a highly sensitive method of detection for *Schistosoma* DNA, with a PCR targeting a 121 bp tandem repeat sequence, has been reported in plasma as well.

2.5 Concluding Remarks

Multiplex real-time PCR provides an extremely sensitive and specific diagnostic alternative for labour-intensive microscopy in clinical laboratory practice. The number of laboratories that implement real-time PCR for the detection of parasitic infections is increasing. This applies in particular to diarrhoea-causing protozoa, which is often combined with a bacterial and a viral enteritis panel, into a total panel. In the western setting, the pathogenic parasites *E. histolytica, G. lamblia* and *Cryptosporidium* make up the vast majority of infections that are detected by microscopy. In travellers, additional parasites are found by microscopy in a small minority of cases, especially in those who have travelled to areas with high risk. Additional diagnostic methods for the detection of parasitic infections that are not included in the "standard" faecal PCR panel can thus be restricted to a selected group of patients. The implementation of molecular diagnostics obviously also calls for quality assurance. These initiatives have been taken and regular assessment schemes are available. Molecular diagnostics of less common parasites and more extensive genetic characterizations are carried out in various research centres. All over the world, including in developing countries, real-time PCR is available in a growing number of research centres and molecular diagnostics of parasitic infections is applied in large-scale epidemiological and fundamental research.

In the future, variations on microarrays and next generation sequencing platforms will undoubtedly become more user-friendly and more applicable in a diagnostic setting. This will enhance the use of DNA-based diagnostics in the microbiological diagnostic laboratory.

References and Recommended Literature

Qvarnström Y, et al. Comparison of real-time PCR protocols for differential laboratory diagnosis of amebiasis. J Clin Microbiol 2005;43:5491–97.

Stensvold CR, et al. The impact of genetic diversity in protozoa on molecular diagnostics. Trends Parasitol 2011;27:53–58.

ten Hove RJ. Molecular detection of intestinal parasites for clinical diagnosis and epidemiology. https://openaccess.leidenuniv.nl/bitstream/1887/14031/18/front.pdf (2009).

Verweij JJ. Application of PCR-based methods for diagnosis of intestinal parasitic infections in the clinical laboratory. Parasitology 2014;141:1863–72.

Verweij JJ, Stensvold CR. Molecular testing for clinical diagnosis and epidemiological investigations of intestinal parasitic infections. Clin Microbiol Rev 2014;27:371–418.

Verweij JJ, et al. Simultaneous detection of Entamoeba histolytica, Giardia lamblia, and Cryptosporidium parvum in fecal samples by using multiplex real-time PCR. J Clin Microbiol 2004;42:1220–23.

Verweij JJ, et al. Real-time PCR for the detection of Dientamoeba fragilis in fecal samples. Mol Cell Probes 2007;21:400–404.

Verweij JJ, et al. Molecular diagnosis of Strongyloides stercoralis in faecal samples using real-time PCR. Trans R Soc Trop Med Hyg 2009;103:342–46.

Wiria AE, et al. Does treatment of intestinal helminth infections influence malaria? Background and methodology of a longitudinal study of clinical, parasitological and immunological parameters in Nangapanda, Flores, Indonesia (ImmunoSPIN Study). BMC Infect Dis 2010;10:77.

Chapter 3
Virology

G.J. Boland

Abstract Molecular techniques have become indispensable in viral diagnostics. Current applications include: (1) The detection of (unknown) viral infections in clinical samples. (2) Quantitative monitoring of viral load. (3) Genotyping of viral strains. (4) Detection of mutations in the viral genome that are associated with viral resistance. Proper sample acquisition and sample transport, as well as accurate DNA or RNA isolation are a prerequisite for reliable test results in molecular diagnostics of viral infections. In-house as well as commercial assays can be used for the amplification of viral DNA or RNA for the detection of viral infections and viral load monitoring. Many virus species consist of several subspecies, genotypes or variants. This molecular variation has to be taken into account when applying molecular diagnostics. More complicated diagnostics for genotyping or the detection of mutations related to therapy failure often rely on sequencing, although for some viral targets commercial assays are available. In this chapter, applications are described in which molecular methods have become the most important form of viral diagnostics. Molecular test results have a direct impact on patient management and as such, results have to be reliable, standardized and reproducible. Therefore, quality control and standardization are important issues!

Keywords Diagnosis · Virus detection · Variation · Genotypes · Resistance

3.1 Introduction

In the past, viral culturing techniques were used for the detection of viruses in clinical samples. Most of these culture methods have been replaced by molecular detection of viruses. The advantages of molecular detection, when compared to virus culture, are the higher diagnostic sensitivity and the much shorter

G.J. Boland (✉)
Division Laboratories and Pharmacy, Medical Microbiology,
UMC Utrecht, Utrecht, The Netherlands
e-mail: G.J.Boland@umcutrecht.nl

E. van Pelt-Verkuil et al. (eds.), *Molecular Diagnostics*,
DOI 10.1007/978-981-10-4511-0_3

time-to-result. A limitation of the molecular detection of viruses, when compared to conventional culture, is that the assays will only detect known viruses that were included in the search panel.

Screening for persisting viruses is usually performed by serological methods. The detection of antibodies is, however, an indirect method to detect an infection and there is usually no correlation with the severity of the disease. Viral load is generally a more accurate indicator of the severity of infection. Molecular techniques for quantitative viral load monitoring are applied to monitor (chronic) infections and to measure the effect of antiviral therapy.

For some viruses it may be important to detect both the presence of a virus and to determine the genotype or strain and specific resistance-associated mutations.

In this chapter applications of molecular detection of viral infection, viral load monitoring and more complex and often also less standardized methods, like genotyping and resistance testing in clinical diagnostics, are discussed.

3.2 Isolation of Viral Nucleic Acids from Clinical Samples

Viral diagnostics are performed on a wide range of clinical materials, including (EDTA) plasma, whole blood, faeces, respiratory or genital swabs, cerebrospinal fluid (liquor), urine, tissue samples or biopsies, broncho-alveolar lavage fluids, etc. Extraction of the viral DNA or RNA is a critical step in the whole procedure of viral diagnostics. For some materials, pre-treatment of the sample is necessary to disrupt the tissue, release the virus or deactivate inhibitory factors. Inhibitory factors for PCR are for instance bile in faeces or haemoglobin in blood samples. Some viruses are not stable and clinical materials have to be processed within several hours to ensure optimal test results. Especially in case of viral load determination, accurate and timely sample processing is essential. This applies for instance to HIV viral load testing in plasma.

Buffers have been developed for some sample types, e.g. faecal samples, which preserve the nucleic acids present in the sample. For swabs, extraction tubes are available in which the swab can be placed directly to preserve the nucleic acids. In case of tissue samples or biopsies, the tissue has to be disrupted to release viral DNA or RNA e.g. by sonication, freezing, special DNA isolation kits for tissue, etc.

Extraction of viral nucleic acids can be done manually or automatically. Commercial manufacturers have developed a number of manual extraction kits. These kits vary in the method they use, the time required for extraction, the samples they are suitable for and the costs. In most methods, DNA or RNA is bound to a solid phase of glass fibres or silica, washed, released and precipitated. Manual extraction requires multiple manipulations and with the increasing number of samples, automated extraction is indispensable in diagnostic laboratories. Automated extraction instruments are manufactured by a number of different companies. By automation and standardization, the recovery of DNA and RNA from clinical samples is consistent and reproducible. More samples can be processed within a shorter period of time as compared with manual isolation, depending on the capacity of the instrument.

3.3 Molecular Diagnostics of Viruses

3.3.1 Viral Detection Methods

Several molecular techniques are used for the detection of viruses. Widely used are target-amplification methods based on the (reverse transcriptase) polymerase chain reaction ((RT-) PCR). Other target-amplification methods are nucleic acid sequence based amplification (NASBA) and transcription-mediated amplification (TMA). NASBA and TMA are isothermal target-amplification techniques and are described in Chap. 7 of Volume 1.

Other techniques are based on signal-amplification. One of these techniques is the branched-DNA (bDNA) assay. This technique is based on binding of the denatured target DNA or RNA on sequence-specific capture probes (bound to a micro well) and capture extenders. Second probes, called label extenders, also bind specifically and include a sequence that also binds to pre-amplifier probes. Multiple amplifiers bind to each pre-amplifier probe, amplifying the signal. This technique is also described in Chap. 7 of Volume 1. The advantages of signal amplifying techniques are that these methods are sensitive, specific, obtain a quantitative result and can be automated. However, these techniques are available for only a few targets (like HIV, hepatitis B and C) and are not suitable for development in your own laboratory.

The most popular technique in many of the virology laboratories is probably real-time PCR (qPCR). Commercially available qPCR based techniques are mainly developed for the most frequently detected targets in diagnostic laboratories. The advantage of qPCR for a molecular diagnostic (virology) laboratory is that the laboratory can develop its own tests for targets for which no CE-marked in vitro diagnostic (IVD) assays are available. qPCR can be used for the qualitative and quantitative detection of viruses. (Multiplex) qPCR can be used for the detection of many different viruses in one assay, for example for the detection of respiratory viruses like influenza virus, para-influenza virus, rhinovirus, respiratory syncytial virus, human metapneumo virus and coronaviruses (respiratory viral panel) in samples from patients with respiratory disease or for the detection of norovirus, rotavirus, enteroviruses and parechovirus (gastrointestinal viral panel) in patients with gastrointestinal disease.

In recent years attention has been paid to workflow optimization and result processing in order to decrease the number of manual actions, diminish deviations and errors and increase standardization and automation.

3.3.2 Primer/Probe Design

The target sequence for the design of primers and probes for the detection of viruses must be unique to the target. Preferably, the PCR primers should be able to identify the target region with high efficiency and specificity (see Vol. 1, Chap. 3). In the development of primers and probes for the detection of viruses, however, the variation of the viral genome has to be taken into account. Especially RNA-viruses, which lack proofreading in the replication of the virus, have high variability. A search in a DNA database, such as the National Center for Biotechnology Information (NCBI) database (http://www.ncbi.nlm.nih.gov/BLAST/), may reveal cross-reactivity with other nucleotide sequences. For clinical applications, it is necessary to validate the assay in the clinical material to which it will be applied, for instance faecal samples, to confirm absence of cross reactivity to normal flora in those specific samples. Mutations in the viral genome situated in the binding region of the primers and probe affect the binding efficiency of the primers and probes and consequently influence the efficiency of the amplification reaction. Both the location of the mutations and the number of mutations are important for the binding of the primers and probes and the amplification efficiency.

For the detection of viruses, primers and probes are preferably designed to anneal to conserved regions of the viral genome. More variable parts of the viral genome usually code for proteins that are targeted by the immune system, such as the envelope proteins or other structural proteins. Obviously, primers and probes designed to target these variable areas will have a greater chance to miss variants.

When multiple targets must be detected in one clinical sample within one PCR run, primers and probes should be used that bind under similar PCR conditions (temperature, concentration, etc.). The design of the primers and probes is important and self- and cross-hybridization has to be avoided. The more PCR reactions are performed in one-tube, the more difficult this will be. When combining different PCR reactions, these factors must be taken into account and comprehensive testing is mandatory when used for patient diagnostics (see Vol. 1, Chap. 5).

3.3.2.1 Mismatch-Tolerance in Primers

Not much literature is available about the effect of mismatches in primers on amplification-efficiency. Mismatch-tolerance is dependent on the type of mismatch (T/A or G/C), the T_m of the primer and the percentage G/C compared to A/T. The effect of mutations on qPCR efficiency also depends on the number and location of the mutations.

Mutations in the target DNA at the 3′-OH end of the primer, the site where the elongation starts, may compromise elongation and PCR efficiency (Vol. 1, Sect. 5.5). This effect of mismatching has been investigated and is directly related to nucleotides involved (Table 3.1). Studies show that mutations at the 3′-part of the primer, where the polymerase starts the elongation, lead to a suboptimal

Table 3.1 Relative amplification efficiency with 3′—primer terminus mismatch

		Primer 3′-terminus nucleotide			
		T	C	G	A
Template 3′—nucleotide	T	0.1–0.01	0.1–0.01	1.0–0.1	1.0
	C	0.1–0.01	<0.01	1.0	1.0–0.1
	G	1.0–0.1	1.0	<0.01	<0.01
	A	1.0	1.0–0.1	<0.01	<0.01

The amount of PCR-product is related to the amount of PCR product with a primer with a perfect match

amplification efficiency and an underestimation of the amount of target. Mismatches A·A (A in the primer and A in the target), A·G, G·A, G·G and C·C lead to an underestimation of the amount of the target by a factor of 100. T·C, C·T, or T·T mismatches result in an underestimation by a factor of 10 to 100 of the target and A·C, C·A, T·G and G·T mismatches have nearly no effect. The inhibitory effect on target amplification decreases when the mutation is moved to the 5′-end of the primer and disappears after 5 nucleotides from the 3′-primer end. The effect of mutations in primers and primer binding sites also depends on the G/C content of the primers and the sequence itself (e.g. the G/C content on the 3′-end of the primer) and the annealing temperature.

The effect of mutations also depends on the amplification enzyme. When using M-MLV/Taq polymerase for reverse-transcriptase, mutations in the reverse-primer binding site have nearly no effect, probably because of the low optimal temperature (37 °C) at which M-MLV is active as reverse transcriptase, compared to AMV (optimal temperature 45 °C). Mutations in the forward-primer binding site seem of less importance when rTth is used as polymerase. Not only the enzymes, but also other factors, like the constituents of the master mix (salt concentration) and hot start, may affect the influence of mutations in the primer binding sites.

In clinical practice, it is desirable to perform multiple PCR reactions on one sample in one PCR run, either as multiplex or in different wells (singleplex). This means that there is a beforehand established annealing temperature and fixed Mg^{2+} concentration. Primers are selected that anneal optimally at this temperature, either by adjusting the length of the primer, by adjusting the sequence or by using nucleotide-analogues (see Vol. 1, Chap. 4).

3.3.2.2 Probe Design

Probe-systems that are mostly used in diagnostics of viral infections with qPCR are hydrolysis probes and hybridisation probes:

- Hydrolysis probes bind to the target sequence during the annealing phase. These probes are designed with a T_m approx. 10 °C higher than the T_m of the primers. During DNA synthesis, in the elongation phase, the hybridized probe is reached

and hydrolysed by the 5′-3′-exonuclease activity of the polymerase. By breaking down the probe, the reporter-fluorochrome (VIC, FAM, TET, etc.) and quencher (TAMRA) are separated and 'fluorescence resonance energy transfer (FRET)' (the signal from the reporter) is released (see Vol. 1, Chap. 3). MGB-or LNA-probes are types of hydrolysis probes that are frequently used in qPCR assays for the diagnosis of viral infections (Vol. 1, Chap. 3). When these probes are designed well, they allow up to three mismatches in the target sequence without much loss of signal. Applications where mismatch-tolerance is needed are for instance qPCR for detection of variable viral targets. However, in virology, there are also applications where it is important that molecular variation is detected, for instance in the detection of viral variants with different pathogenicity or in the detection of mutations that are associated with resistance to antiviral medication. So, depending on the purpose of the assay, more or less mismatches can be tolerated.
- Hybridisation probes used in viral diagnostics can be used as molecular 'beacons'. Molecular 'beacons' are long probes with a loop structure, which contains the part of the probe that binds to the target region. Attached to the probe are two G-C rich, complementary regions that create the hairpin structure. One side of the probe is labelled with the fluorescent dye, the other with the quencher. These probes release a fluorescent signal upon hybridization to the target DNA and are not hydrolysed during the elongation phase in the PCR reaction. Molecular 'beacon' probes can be designed (T_m of the loop compared to the T_m of the stem) to be very sensitive to detect mutations, for example mutations that are associated with viral resistance. Other hybridization probes are double stranded probes containing a long strand (>40 nt) labelled with the reporter fluorochrome and a short strand homologous to the reporter probe (<16 nt), which is labelled with the quencher molecule. During elongation, the reporter probe will preferably bind to the target DNA and the probe with the quencher will be released, thereby releasing the signal. This method allows detection of the signal at much lower temperatures (until 37 °C) and is much less sensitive to mutations in the target sequence without affecting the PCR efficiency. However, these probes are more difficult to design compared to hydrolysis probes.

3.4 Molecular Techniques for the Detection of Viral Variation

It is not only important to detect a virus, but often the analysis of genotypic characteristics of the virus is also important. Within one species, there are often strains with different characteristics and different pathogenicity, in need of other treatment strategies. An example is the group of enteroviruses. The genus enterovirus consists of the species echovirus, rhinovirus, poliovirus, enterovirus and coxsackievirus. Furthermore, these species consist of several subspecies and strains.

In viral diagnostics, it is important to distinguish, for instance, the pathogenic poliovirus from other species within the genus of enteroviruses, but it may also be relevant to type viruses within the species. This can be done to obtain information on the relationship between different isolates in case of outbreaks, to identify strains with increased virulence or for epidemiological purposes. Another example is human papillomavirus (HPV). HPV is associated with cervical cancer. Within this genus there are high-risk HPV (hrHPV) strains, (e.g. HPV 16, 18, 31, 33, 45, 53, 66 and 68), probably hrHPV strains (e.g. HPV 26) and low-risk HPV strains (e.g. 40, 42, 72, 81 etc.). Distinguishing these (probably) hrHPV strains from the low-risk strains is essential for clinical diagnostics.

Another application is the molecular detection of resistance to antiviral medication. This field is becoming more important with the development of new antiviral medication and the occurrence of resistance variants of a variety viruses. Examples of viruses for which resistance to antiviral medication is important are HIV, hepatitis B, hepatitis C, herpes simplex virus type I and II and the influenza virus.

A few techniques are available to detect viral variation. The most frequently used and also the most informative one, is sequencing (part of) the viral genome. Sequencing can be used for several purposes, like genotyping (detection of variation) and the detection of mutations associated with resistance. A disadvantage is, however, that it is also the most time-consuming procedure. With the development of sequencing technology, it has become less expensive and more feasible for diagnostic laboratories (see Vol. 1, Chap. 8).

Sequences of viral genomes can be uploaded into online databases (e.g. NCBI) to determine the genotype of the virus. For some viruses, tools are available on the internet, e.g. the NCBI genotyping tool, Geno2Pheno or HIV-GRADE, either for genotyping or to quickly determine mutations associated with antiviral resistance. These tools are not available for all viruses. For other viruses, sequences can be uploaded into online databases and compared to viral sequences retrieved from GenBank or the Los Alamos database. Programs like NCBI blast or SeqScape can be used to compare sequences to detect mutations.

In some cases, information of only a selected part of the genome or of only a few mutations is necessary for subtyping or detection of antiviral resistance. For some applications, for instance genotyping of human papillomavirus (HPV), hepatitis B and hepatitis C, or for the detection of drug-resistant variants for hepatitis B, hybridization assays have been developed (LiPA). These assays consist of a qPCR combined with a hybridization assay (Vol. 1, Chaps. 4 and 5). The hybridization assay is performed on nitrocellulose strips that contain lines loaded with genotype-specific oligonucleotides. To detect drug-resistant variants, the strips contain wild type and mutant-specific oligonucleotide probes. The advantage of this method, compared to sequencing, is that more samples can be processed in the same time and that the evaluation of test results is less time-consuming. However, an obvious disadvantage is that mutations appearing outside the oligonucleotides used in the assay are not detected and newly emerged variants are not distinguished. Furthermore, natural variation of the virus is not visible.

Sequencing provides more information about the viral genome. Direct sequencing of a PCR product with the Sanger method does not yield information about variants that are present in low percentages in the total viral population. Variants that are present in less than $\sim$25% of the total viral population are not detected. Next generation sequencing methods are more sensitive for the detection of minor viral populations. LiPA-assays also tend to be better in the detection of minority strains when compared to Sanger-sequencing methods.

3.5 Clinical Applications of Molecular Diagnostics

3.5.1 Molecular Detection of Viral Infections

In clinical diagnostics, the question is often: "which virus is causing this disease?" When using molecular diagnostics to answer this question, the symptoms of the disease and patient characteristics also give directions to the most likely etiological agents. For instance, respiratory diseases can be caused by a number of viruses like influenza, parainfluenza, adenovirus, coronaviruses, rhinoviruses and human metapneumo virus, but in children, the respiratory syncytial virus is also a common cause. Infections of the upper respiratory tract may have different causes than infections of the lower respiratory tract. Additionally, patient-related factors like asthma or other chronic diseases, or medication, may influence the type and seriousness of the infection. Most diagnostic laboratories and diagnostic companies have developed specific respiratory panels to detect the most common causes of respiratory infections. Also for gastroenteritis and meningitis, there are multiple causes and usually a panel of viruses is tested. The most frequent cause of viral meningitis are enteroviruses, but occasionally, viral meningitis can be caused by other viruses like the herpes simplex virus, the mumps virus or arboviruses, which may require other precautions and treatment. Examples of viral causes of diseases are shown in Table 3.2.

The advantage of applying these panels is that all major causes of the infection are included and can be tested in a single assay. A disadvantage can be that the infection of a specific patient is not included in the assay or that the patient may have a variant of a specific virus that is not detected with the assay that is applied.

3.5.2 Quantitative Detection of Viruses—Viral Load

Molecular techniques have made it possible to obtain a quantitative or semi quantitative result that can be related to a viral load. To obtain a reliable quantitative

Table 3.2 Common viral causes of diseases*

Clinical characteristic	Viral causes	
Upper respiratory tract (Mostly viral causes)	Rhinovirus (RNA)	>100 types, early and late
	Coronavirus (RNA)	summer/autumn
	Adenovirus (DNA)	5 types, winter and spring season
	RSV** (RNA)	>50 types, winter and spring
	Enteroviruses (RNA)	2 types, winter season, children
	Influenza (RNA)	>90 types, summer and autumn
	Parainfluenza (RNA)	Winter and spring
	Human parechovirus (RNA)	Autumn, winter and spring 14 types
	Other: EBV/HSV/MERS and SARS coronavirus	
Lower respiratory tract (Mostly bacterial causes, only viral causes mentioned)	Influenza RSV (children)	Lower respiratory infections with viral causes may occur in immune suppressed persons or as superinfections
	Other: coronavirus, rhinoviruses, parainfluenzavirus, adenovirus, VZV	
Gastroenteritis (also bacterial and parasitic causes)	Norovirus (RNA)	Winter/outbreaks, mostly adults
	Rotavirus (RNA)	Most common in children
	Astrovirus (RNA)	Children and adults
	Sapovirus (RNA)	Mostly adults
Erythema/skin disease	Mumps (RNA)	Erythema
	Rubella (RNA)	Erythema
	Measles (RNA)	Erythema, measles
	Parvovirus B19 (RNA)	Erythema
	Herpes simplex (DNA)	Herpes virus, 2 types, vesicles
	VZV (DNA)	Herpes virus, vesicles
	HHV-6 (DNA)	Herpes virus, vesicles
Hepatitis	Hepatitis A (RNA)	Acute hepatitis
	Hepatitis B (DNA)	Acute and chronic
	Hepatitis C (RNA)	Mostly chronic
	Hepatitis D (RNA)	Subviral particle, only as coinfection with hepatitis B
	Hepatitis E (RNA)	Mostly acute
	Other: EBV, CMV	Not specific hepatitis virus but may cause hepatitis occasionally
Meningitis	Enteroviruses	
	Parechovirus	
	JC virus	
	Other: mumps, measles, poliovirus, HIV, herpesviruses	
Other viral causes	Parvovirus	Joint pain
	Dengue virus	Travelers, fever and joint pain
	HIV	Immunodeficiency, sexual transmission
	Yellow fever virus	Travelers
	West-Nile virus	Travelers

*Table may not be complete and changes may occur as new viruses may evolve
**Respiratory syncytial virus

test result that can be compared with other test results, the amount (or input) of the clinical sample has to be standardized. Preferably, the test result relates to an international standard to calculate the viral load. Viral load is used to monitor herpes virus reactivation after transplantation and in some chronic viral infections like hepatitis B, hepatitis C and HIV. Since virus detection and viral monitoring have become essential in patient care, it is necessary that these techniques are performed frequently and on a regular basis in molecular diagnostic laboratories. Automation of the workflow from nucleic acid isolation, amplification, detection and processing of the results is essential for laboratories to maintain a high degree of standardization, to ensure that test results are reproducible.

There are three major reasons for quantitative detection of viral DNA or RNA. The first reason is that the amount of virus ('viral load'), which is usually reflected in the amount of viral DNA or RNA present, can be an indication of the seriousness of the infection. This was first demonstrated for HIV by using a quantitative HIV-1 bDNA technique. An increasing amount of HIV-1 RNA genomes in plasma was directly correlated with a decrease in CD4+ T-cells, with disease progression and HIV-1 induced mortality. Since these findings were published, HIV-1 viral load monitoring is a major factor in the decision for antiviral treatment.

Since then, quantitative viral load determination is implemented in patient monitoring for other viral infections as well. Most people are infected with cytomegalovirus and Epstein Barr virus. These viruses usually do not cause problems, but occasionally cause viremia (virus detectable in blood) without symptoms. However, these viruses can cause major problems in patients after stem cell transplantation. Treatment against these viruses is installed when a certain viral load is reached. For EBV, it was shown that a viral load $>10^3$ copies/ml is predictive for the development of post-transplant lymphoproliferative disease. By starting antiviral therapy when this viral load is reached, but before symptoms appear (pre-emptive therapy), it was shown that the EBV-related mortality significantly reduced in these patients.

A second reason for quantification of the viral DNA or RNA is that the risk of viral transmission is directly related to the amount of virus present. Vertical transmission (from mother to child during pregnancy or at birth) is an example. Vertical transmission of HIV can be prevented either by decreasing the viral load during pregnancy by antiviral therapy, or by performing a caesarian section. For hepatitis B virus, it is also shown that the risk of vertical transmission is directly related to the viral load of the mother. Infection can be prevented by immediate active and passive vaccination of the newborn. However, the risk of vertical transmission is still present if the mother has a high viral load ($>2 \times 10^8$ IU/ml). In that case, it is also recommended to start antiviral treatment in the last trimester of the pregnancy.

Alternatively, in case of needle stick incidents, the viral load is directly related to the chance of transmission of the virus. Needle stick accidents is a collective name for incidents in which health care workers accidently come into contact with patient fluids that may be contaminated with hepatitis B, hepatitis C or HIV. Accidental blood contact occurs frequently in hospitals. Measures taken after accidental contact comprise monitoring of viral load in both the victim of the needle stick accident and the patient and/or preventive antiviral treatment. Vice versa, health care workers may also infect patients. Several countries have taken measures to prevent transmission of viral infections to patients. For hepatitis B, several cases of transmission from health care workers to patients are known. Physicians with a high hepatitis B viral load are forbidden to perform medical procedures that pose a risk for patients in most countries.

A third reason for the quantification of viral DNA or RNA is monitoring the effect of antiviral treatment. The ultimate goal of antiviral treatment is to prevent damage to the host by the infection. Antiviral treatment is installed to reduce the viral load and to prevent this damage. Applications to monitor the effect of antiviral treatment are being used for HIV, hepatitis B, hepatitis C and herpes viruses, like cytomegalovirus and Epstein Barr virus. The primary goal is to decrease the viral load to an undetectable limit. Monitoring of the decrease in viral load directly after starting therapy can have a predictive value for treatment success. For some viruses, the absence of a certain decline in viral load may be an indication to stop or to switch treatment.

Some conditions influence the reliability of the quantification of a viral load. First, the sampling procedure and the handling of samples may influence the test results. For example, varicella zoster virus is a cell-associated virus. In vesicle fluid, the virus load can be low and may result in a false-negative PCR test result. Scraping of the bottom of the vesicle, where most virus-infected cells are present, may result in a very high viral load. Second, the viral load may not fluctuate too much in relation to the time that is needed to generate a quantitative test result. In acute viral infections, the virus load increases quickly to very high levels and decreases after that. For many acute viral respiratory infections it is not possible to show a relation between the viral load and the seriousness of the respiratory disease. Third, variation in virus load that is caused by the procedure has to be avoided. There are many factors that influence the outcome of the viral load: sample handling and freeze thawing, pipetting deviations, sample viscosity, batch-differences in reagent preparation and PCR-kits, etc. To obtain reliable quantitative test results, these factors have to be avoided.

Automation in quantitative PCR reduces analytical variation. In some quantitative PCR assays, viral load determinations are calibrated using an international reference strain, which makes comparison of viral loads between laboratories and between assays more reliable.

3.5.3 Genotyping and Genotypic Determination of Viral Resistance

Due to a combination of sloppy replication, immune pressure and other factors, viruses have evolved into different geno-or subtypes. Clinical features of genotypes may differ and sometimes even influence the choice of therapy. Viral resistance can occur during treatment and determines the course of the infection and the choice of medication. Often, only small variations, or even single nucleotide point mutations (SNP), may have a great impact on the pathogenicity, replicative activity and/or susceptibility for treatment. Mutations are usually represented as the amino acid that is replaced and the position, e.g. Q151 M is a mutation in the reverse transcriptase gene of HIV, in which the glutamine at position 151 in the protein is replaced by a methionine. A few examples are discussed in the following subsections.

3.5.3.1 HIV-1

Antiviral treatment for HIV consists of a combination of antiviral drugs:

- Nucleotide/nucleoside reverse transcriptase inhibitors (NRTI's)
- Non-nucleotide reverse transcriptase inhibitors (NNRTI's)
- Protease inhibitors (PI's)
- Integrase inhibitors (INSTI's)
- Fusion inhibitors
- CCR5 co-receptor antagonist (maraviroc)
- Pharmacokinetic booster agents like ritonavir.

Genotypic resistance testing is usually performed on the reverse transcriptase and protease genes of HIV and it is performed in case of suboptimal viral reduction and to assist in the selection of active drugs. Mutations that lead to high-level reduced susceptibility to integrase inhibitors have been described. Pre-treatment resistance testing is advised, due to the risk of transmission of resistant virus strains.

Some mutations are able to individually induce reduced susceptibility to antiviral drugs, whereas some mutations invoke resistance only in combination with other mutations. For anti-HIV medication, the terms low- or high-genetic barrier are often used to indicate whether resistance can evolve easily or requires more than one mutation. Resistance is usually tested by sequencing part of the genome. Various online tools can assist the laboratory workers with the analysis of genotyping test results. The international AIDS society (USA) maintains a list of resistance-associated mutations (http://www.iasusa.org/resistance_mutations, login required) and Stanford University maintains a HIV Drug resistance database (http://hivdb.stanford.edu).

3.5.3.2 Hepatitis B

The hepatitis B virus consists of genotypes A through H. The small differences between these genotypes are responsible for the differences in the risk of the development of liver carcinoma and the response to treatment. Treatment of hepatitis B with antiviral medication aims at long-term suppression of the hepatitis B virus and at reducing related morbidity and mortality. Thus far, the differences in genotypes have only been shown to have minor consequences for treatment strategies, but they are of interest for predictions. For instance, hepatitis B genotypes A and B seem to respond relatively well to treatment with Interferon-alpha, compared to, for instance, genotype C and D. A physician tends to preferably treat genotypes A or B with Interferon. Treatment with Interferon aims at improving the antiviral response of the immune system, at reducing viral susceptibility of cells by enhancing Major Histocompatibility Complex (MHC)-antigen expression and at inducing cellular proteins that are important in the natural defense of cells against viruses. This leads to immune control of the chronic hepatitis B infection.

Direct-acting antivirals (DAA's, nucleotide-or nucleoside-analogues) that are available for the treatment of an infection with the hepatitis B virus are lamivudine, adefovir, telbivudine, tenofovir and entecavir. One mutation, called the YMDD motive, is associated with resistance to lamivudine. The YMDD motive is a conserved sequence in the RNA-dependent DNA polymerase gene of hepatitis B virus consisting of tyrosine (Y), methionine (M), and two aspartates (DD). Replacement of the methionine by isoleucine (YIDD) or valine (YVDD) leads to resistance. Since the polymerase and hepatitis B surface antigen genes overlap, replacement of nucleotides may also lead to changes in the surface antigen. Therefore, the mutations also have an impact on the production and secretion of the HBsAg. Lamivudine has long been the only treatment option for hepatitis B, but now has been replaced by other more potent antivirals, like entecavir and tenofovir. Entecavir, however, is less effective in patients who have been pretreated with lamivudine and who carry the YMDD variant of hepatitis B. Genotyping and resistance testing of hepatitis B can be performed by sequencing and comparing the sequence to reference strains, but LiPA assays have also been developed for genotyping and for the detection of mutations associated with resistance.

3.5.3.3 Hepatitis C

As an alternative to HIV and hepatitis B treatment, which aims at long-term suppression of the virus, the treatment of the hepatitis C virus aims to totally clear the virus from the host. In the past decades, hepatitis C was treated with a combination of interferon-alpha and ribavirin. Nowadays, an increasing number of direct-acting antivirals (DAA's) are available. Several viral proteins involved in the HCV lifecycle, such as the non-structural (NS) 3/4A serine protease, the NS5B RNA-dependent RNA polymerase (RdRp) and the NS5A protein, are targets for these drugs. NS5A plays a role in the formation of a replication complex of the virus and in the release of new viral particles. The Hepatitis C virus has 7 major genotypes, all with different prevalence and different responses to treatment. Genotypes 1, 2, 3 and 4 are most prevalent in western countries. Genotyping can be performed with a LiPA assay, but sequence-based assays have also been developed. First-generation protease-inhibitors telaprevir and boceprevir are active against hepatitis C genotype 1, but they are less effective and therefore not registered for other genotypes. New NS3-protease-inhibitors are registered for treatment of genotype 1 and 4. Polymerase inhibitors and NS5A inhibitors show less genotypic specificity. Since treatment with one antiviral is often ineffective, combination therapy is always given. The choice of medicine largely depends on the hepatitis C genotype.

Antiviral resistance does not appear to be a serious problem in treatment with NS5B polymerase-inhibitors. For NS5A-inhibitors, some clinically relevant mutations have been described, like M28T, Q30E/H/R, L31 M/V and Y93H. Drug resistance to protease-inhibitors seems more complex. Q80 K is a mutation that is related to resistance to simeprevir, but only in genotype 1a-infected patients (Fig. 3.1).

R155 K/T is a mutation that appears to invoke resistance to almost all protease-inhibitors, including to protease-inhibitors still in development in genotype 1a-infected patients. The D168 V mutation seems to be prevalent mostly in genotype 1b-infected patients and is related to resistance. Mutations associated with resistance may be present before treatment or develop during treatment. Pretreatment sequence analysis and monitoring of response to antiviral treatment will give more information that may be of help in future therapy development.

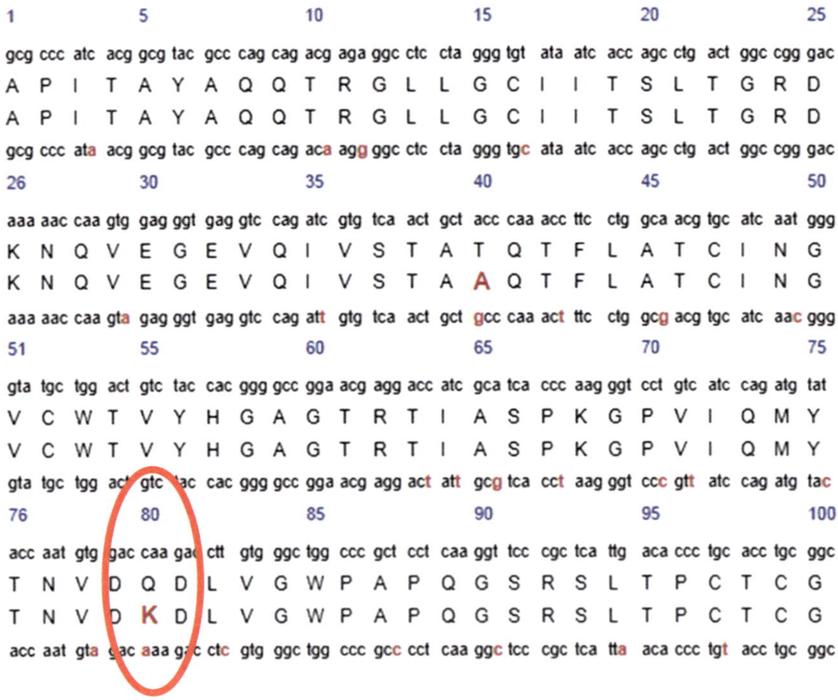

Fig. 3.1 Alignment of a part of HCV protease with reference strain HCV gt 1a N77—translation of nucleotides to amino acids

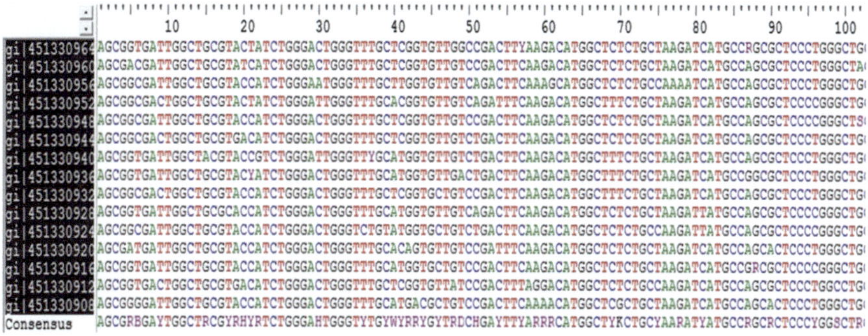

Fig. 3.2 HCV NS5A sequence variation within HCV genotype 1a

Detection of drug-resistant variants is done with sequence-based assays. Sequence variation between the genotypes of hepatitis C is large and makes it difficult to design primers for amplification and sequencing of the protease-, NS5A or polymerase genes of hepatitis C. In Fig. 3.2 it is shown that even within a hepatitis C genotype and subtype, there is considerable sequence variation (Fig. 3.2).

3.5.3.4 Herpes Viruses

Herpes simplex virus 1 and 2 (HSV 1 and 2), varicella zoster virus (VZV) and cytomegalovirus (CMV) belong to the herpes virus family, which can be treated with specific antiviral drugs. Resistance has been described for this family of viruses. Sequencing part of the genome is necessary to detect involved mutations.

HSV 1 and 2 cause lesions in the mouth that are usually not harmful. However, these lesions can become severe in immune suppressed patients and systemic illness may occur. For HSV 1 and 2, 7–14% of patients that receive treatment with acyclovir, or similar medications like valaciclovir and famciclovir, become resistant. Up to 95% of resistance to acyclovir is associated with mutations in the thymidine-kinase gene (UL23). Viral thymidine-kinase phosphorylates acyclovir, which is a necessary step in the antiviral mechanism of acyclovir, before it can act as a nucleotide analogue to disturb viral DNA synthesis. Acyclovir-resistant variants may still be sensitive to foscarnet. Foscarnet blocks DNA polymerase, thereby inhibiting viral DNA synthesis. Mutations in UL30 (polymerase-gene) are associated with resistance to foscarnet. In both genes, many polymorphisms may be present. Resistance to acyclovir is often associated with insertions or deletions in the thymidine-kinase gene. Also, some single mutations are associated with resistance. Thorough analysis of the gene sequences, comparisons to reference strains and to mutations found in literature are necessary for analysis.

The same genes (thymidine kinase and polymerase) are involved in resistance of VZV to antiviral therapy. The first infection with varicella zoster virus gives chickenpox. Varicella zoster virus remains latently present in neural axons of the central nervous system and can reactivate, causing shingles. Treatment is similar to herpes simplex virus treatment, with acyclovir or foscarnet. Resistance testing is also performed by sequencing (part of) the thymidine kinase and polymerase genes and comparing to reference sequences.

First choice of treatment for CMV is ganciclovir. Ganciclovir is a guanosine analogue that exerts its antiviral effect by inhibiting CMV DNA polymerase. Ganciclovir is phosphorylated by a viral phosphotransferase coded on UL97. The CMV DNA polymerase is a product of UL54. Mutations in both genes can lead to resistance. However, mutations in UL97 are significantly more common than mutations in UL54. Mutations in UL54 are associated with higher levels of resistance to ganciclovir or cross-resistance to cidofovir or foscarnet.

3.5.3.5 Influenza Virus

Influenza is a member of the family of Orthomyxoviridae, of which the virus species influenza A, influenza B and influenza C can cause our classical 'flu'. The influenza A species consists of different strains, designated according to the structure of their haemagglutinin (HA) and neuraminidase (NA) genes. These are the most variable viral proteins and are involved in attachment to the target cell and fusion with the endosomal membrane. Influenza viruses have a segmented RNA-genome (8 segments), the NA and HA-coding sequences are on different segments and can be exchanged. HA is encoded on segment nr 4 and NA on segment nr 6. Within influenza A, there are at least 16 HA-encoding segments known, of which segments numbered H1–H5 are known as human variants, and 9 NA-variants. Recombination of influenza variants may lead to new combinations of H- and N-genes (antigenic shift), whereas mutations in the H- and N-genes lead to new variants with slightly different properties (antigenic drift). This antigenic drift causes that for instance two A(H3N2) strains may have slightly different antigenic properties. Antigenic drift of less pathogenic viral strains, however, may lead to more pathogenic strains. Usually, every winter season there are one or a few dominant influenza strains active, often one or two A-strains and one B-strain. Within these strains, different isolates can be distinguished.

Compared to influenza A, influenza B is genetically less divers, infection is nearly only restricted to humans and influenza B is thought to be less pathogenic. Influenza C is far less common than influenza A and B, but can cause severe illness and is incidentally associated with outbreaks.

In viral diagnostics, a pan-influenza PCR is used, with primers that detect influenza species A (Matrix gene segment) and a different PCR to detect influenza B. To monitor influenza activity during the winter season and to estimate the effectiveness of the vaccine, viral strains are typed for H and N-gene segments every year. Typing can be performed by reverse transcriptase qPCR with type-specific primers and probes. Some commercial assays are also available. Strain typing is used to predict upcoming strains and to determine the strains that need to be included in the vaccine. The virus mutates from year to year, making existing vaccines ineffective.

For more specific information, the virus has to be sequenced. The whole genome consists of around 2,800,000 nucleotides. Within influenza strains, different isolates can be distinguished, revealing the highly variable genetic structure of this virus. In patient care, the strain type is related to sensitivity for antiviral treatment. Most influenza A and B strains are sensitive to oseltamivir and zanamivir, two neuraminidase inhibitors. Occasionally, resistant strains are found, sometimes related to long-term treatment of influenza infection in immunocompromised patients. Resistance to oseltamivir has also been reported during the 2007–2008-winter season, due to a mutation on position 275 of the neuraminidase gene, resulting in the substitution of a histidine to a tyrosine (H275Y). Strains that have this mutation, however, are still susceptible to zanamivir.

3.6 Conclusions

Molecular diagnostics have become indispensable in the diagnosis of viral infections. qPCR techniques are not only used for the detection of viruses, but also for viral load determination. Whereas chronic viral infections are usually detected by serological methods to test immunity, molecular techniques have become important in the monitoring of chronic viral infections and in the response to antiviral treatment.

By applying molecular techniques we have learned more about viral variation and the different genotypes that exist within virus species. This variation has clinical impact. For some viruses, specific mutations in the viral genome can be related to resistance to antiviral medication. More rapid and cheaper molecular techniques, like next generation sequencing, will be of aid in exploring this field more extensively.

Recommended Literature

Annels NE, et al. Management of Epstein-Barr Virus (EBV) reactivation after allogeneic stem cell transplantation by simultaneous analysis of EBV DNA load and EBV-specific t cell reconstitution. Clin Infect Dis. 2006;42(12):1743–8.

Christopherson C, et al. The effects of internal primer-template mismatches on RT-PCR: HIV-1 model studies. Nucleic Acids Res. 1997;25(3):654–8.

Espy MJ, et al. Real-time PCR in clinical microbiology: applications for routine laboratory testing. Clin Microbiol Rev. 2006;19(1):165–256.

Ghedin E, et al. Large-scale sequencing of human influenza reveals the dynamic nature of viral genome evolution. Nature. 2005;437:1162–6.

MacKaya WG, et al. Molecular detection and typing of influenza viruses: are we ready for an influenza pandemic? J Clin Virol. 2008;42:194–7.

Schutten M, et al. Clinical utility of viral quantification as a tool for disease monitoring. Expert Rev Mol Diagn. 2001;1(2):153–62.

Smith DB, et al. Expanded classification of hepatitis C virus into 7 genotypes and 67 subtypes: updated criteria and genotype assignment web resource. Hepatology. 2014;59(1):318–27. doi:10.1002/hep.26744.

Stuyver L, et al. Typing of hepatitis C virus isolates and characterization of new subtypes using a line probe assay. J Gen Virol. 1993;74(6):1093–102.

Vercauteren J, Vandamme AM. Algorithms for the interpretation of HIV-1 genotypic drug resistance information. Antiviral Res. 2006;71(2–3):335–42.

WHO information for molecular diagnosis of influenza virus—update May 2015. Available at: http://www.who.int/influenza/gisrs_laboratory/molecular_diagnosis_influenza_virus_humans_update_201403rev201505.pdf?ua=1. Accessed 28 Jul 2015.

Chapter 4
Mycology

Jochem B. Buil, Jan Zoll, Paul E. Verweij, Willem J.G. Melchers and A.M.C. Bergmans

Abstract Fungal infections/mycoses are an important cause of morbidity and mortality, especially in patients with an impaired immune system. The clinical diagnosis relies mainly on conventional methods like direct microscopy, culture, radiographic imaging and the evaluation of clinical symptoms. Identification of the fungal pathogen is possible when a positive culture is available, but the diagnostic sensitivity of fungal culture is very low. Molecular assays have been developed to aid in the identification of the fungi when cultures remain negative. These assays can be used to identify the fungi when fungal elements are seen in clinical material. Recently, the more sensitive genus- and species-specific qPCR assays are increasingly being used to aid in the rapid diagnosis of fungal infections. Many PCR assays for the direct detection and identification of dermatophytes in clinical material have been developed during the last decade. The introduction of real-time PCR assays in routine clinical laboratories has speeded up dermatophytosis diagnostics tremendously from turnaround times of two to four weeks (culture), to less than one day. It has also resulted in a huge reduction in hands-on time. Another major advantage of (real-time) PCR is the superior sensitivity as compared to dermatophyte culture. The development of molecular diagnostics has greatly improved the possibilities for faster, more guided treatment of dermatologic disorders, resulting in a significant decrease of adverse effects of antifungals.

Keywords Panfungal · Ribosome · rRNA · Fungal · Immunocompromised · Real-time PCR · Melting curve analysis · Dermatophytes · Trichophyton · Microsporum · Epidermophyton

---

J.B. Buil (✉) · J. Zoll · P.E. Verweij · W.J.G. Melchers
Department of Medical Microbiology, Radboud University Medical Centre, Nijmegen, The Netherlands
e-mail: Jochem.Buil@radboudumc.nl

A.M.C. Bergmans (✉)
Medical Molecular Microbiologist, Laboratory of Medical Microbiology, Bravis Hospital, Roosendaal, The Netherlands
e-mail: a.bergmans@bravis.nl

E. van Pelt-Verkuil et al. (eds.), *Molecular Diagnostics*,
DOI 10.1007/978-981-10-4511-0_4

4.1 Introduction

Laboratory diagnostics of fungal infections, both invasive and dermatophytosis, mainly relies on time-consuming (two to four weeks), labour-intensive and not very specific and sensitive culture methods. To circumvent these issues, many qPCR assays have been developed and introduced to aid in rapid and reliable diagnosis of fungal infections. In this chapter, the major applications of these molecular methods are discussed.

4.2 Molecular detection of invasive fungal infections

Jochem B. Buil, Jan Zoll, Paul E. Verweij and Willem J.G. Melchers

4.2.1 Invasive Fungal Infections

More than 500 species of fungi have been described that can infect humans. Most of these fungi are not harmful for people with an intact immune system. However, immunocompromised patients are susceptible for the so called opportunistic fungal infections. Infections caused by fungi are called mycoses. Mycoses can be caused by yeasts, moulds, and dimorphic fungi. Yeasts are unicellular fungi that duplicate by budding; a bud or outgrowth is formed on the parental cell which will eventually separate to form a daughter cell. Examples of yeast-like fungi causing human disease are species belonging to the genus *Candida*, *Cryptococcus*, and *Pneumocystis jiroveci*. Moulds grow through mycelia, a network composed by multicellular filaments called hyphae. Medically important moulds include species of *Aspergillus*, the Mucorales, *Fusarium* and *Scedosporium*. Several fungi can exist in both the yeast-like form and as moulds. These are called dimorphic fungi. Examples are *Penicillium marneffei, Coccidioides, Histoplasma* and *Blastomyces*. These latter pathogens belong to the endemic fungi, which are fungi with a specific geographical distribution.

4.2.1.1 Diagnosis of Invasive Fungal Infection

The number of patients at risk for invasive fungal disease has increased during recent decades, due to changes in the management of haematological malignancies, solid organ transplantation, and increased use of biologicals. Consequently, the incidence of invasive fungal disease has increased as well. However, the incidence of invasive infections due to *Candida albicans* has been decreased due to the use of fluconazole prophylaxis. Anti-mould prophylaxis has also been proven to be effective in certain patient groups at risk for opportunistic mould infections. Mortality rates of opportunistic invasive fungal infections still approach 50%. Antifungal agents have varying spectrum of activity. Therefore, rapid identification of the pathogen is critical as timely initiation of antifungal therapy is an important component for the outcome of the patient.

The clinical diagnosis of invasive fungal disease remains difficult and relies on direct microscopy of histopathological samples, culture, radiological imaging, all in addition to the evaluation of clinical symptoms. With the primary reason of classifying patients for inclusion in studies and to aid in the communication between clinicians and researchers, definitions for invasive fungal disease (excluding Pneumocystis pneumonia) have been established by a group of specialists: The European Organization for Research and Treatment of Cancer (EORTC) and the Mycoces Study Group (MSG) Consensus group. The EORTC/MSG defined invasive fungal disease with three levels of probability. The categories are defined as "proven", "probable" and "possible" invasive fungal disease. Notably, PCR methods are not included in these criteria for invasive fungal disease, due to lack of clinically validated methods. Therefore molecular methods to detect pathogenic fungi are not being routinely implemented in many microbiology laboratories.

In addition, several guidelines regarding the diagnosis and management of invasive fungal infections have been published during the last several years by organisations like the European Society of Clinical Microbiology and Infectious Diseases (ESCMID), European Conference on Infections in Leukaemia (ECIL) and Infectious Diseases Society of America (IDSA). The recommendations formulated in these guidelines regarding molecular detection of fungal infections are outlined in the following paragraphs. Most older guidelines do not make recommendations for fungal diagnosis or the use of PCR methods. However, recent guidelines have incorporated such recommendations. An overview of recommendations for molecular diagnosis of specific fungal diseases in the various guidelines is shown in Tables 4.1, 4.2, 4.3 and 4.4.

4.2.1.2 Conventional Species Identification

Conventional species identification of moulds relies greatly on careful examination of fungal elements in tissue and the macroscopic and microscopic morphology in culture. Direct microscopy of clinical specimens can only be used for the general discrimination between yeasts, septated moulds and non-septated moulds. Optical brighteners such as Blankophor P help to recognize fungal structures as opposed to much used bacterial stains including the Gram stain. Septated moulds include, but are not limited to, *Aspergillus*, *Scedosporium* and *Fusarium* species. Mucorales can be discriminated from the *Aspergillus* like moulds as their hyphae are broad and non-septated (Fig. 4.2). Further identification of moulds based on microscopic morphology is not possible. Therefore, fungal culture is required for genus and species identification. However, the diagnostic sensitivity (see Vol. 1, Chap. 6) of fungal culture is very low for detection of opportunistic fungi. Even when hyphae are clearly visualized in tissue or respiratory samples, fungal cultures often remain negative. Moreover, even if fungal growth is present, it will take several days before most fungi start growing and even more days before the fungus is sufficiently mature to allow microscopic identification, thus compromising the contribution of conventional diagnostics to guiding antifungal therapy. Finally, many fungi cannot be identified to the species level by conventional methods alone, requiring additional molecular analysis even in culture-positive patients. As several of these so called cryptic species show distinct antifungal susceptibility patterns, fast and reliable molecular methods are essential for clinical management.

4.2.1.3 Molecular Detection Methods

Although molecular detection methods are not yet included in the EORTC/MSG criteria for invasive fungal disease, several studies have shown the utility of PCR for the detection of invasive moulds in human samples. As timely initiation of appropriate antifungal therapy is critical for clinical outcome, early identification of fungi using molecular methods would potentially improve outcome, particularly when the fungal culture is negative. One major problem encountered in molecular detection is the wide variety in procedures regarding the use of the most commonly used technique, the PCR. The variation involves all steps of the procedure including sample type, sample volume, DNA extraction methods, amplification protocols and detection. Therefore, a group of, mainly European, experts founded the European Aspergillus PCR Initiative (EAPCRI). This group collaborates to standardize methods used to detect *Aspergillus* DNA and to validate these methods in clinical trials, with the aim to incorporate molecular methods in future guidelines for the detection and diagnosis of invasive fungal disease.

As susceptible hosts are at risk for infection due to a range of fungal pathogens, one approach to molecular diagnosis is a so-called pan-fungal assay. An optimal pan-fungal assay has the ability to amplify the DNA of all fungal species. As moulds are omnipresent in the environment, using such a non-genus-specific and non-species-specific assay requires a closed PCR system, to exclude contamination by environmental fungal DNA. Most of the pan-fungal assays target parts of the multi-copy rRNA genome of fungi: the 18S, ITS-1, 5.8S, ITS-2 and/or 28S subunits (Fig. 4.1). These parts contain highly conserved regions for all fungal genera as well as highly variable regions, allowing genus and species identification. Several universal primers are available. These primers are designed to target the conserved parts of the fungal rRNA genome. Following amplification, sequence analysis of the variable regions within the rRNA genome enables species identification. Since a pan-fungal assay requires sequence analysis, it is time-consuming and expensive, which are two important disadvantages. Furthermore, the amplicon of a pan-fungal assay is generally larger than 300 base pairs. Due to this size, diagnostic sensitivity is limited and the technique cannot be used to exclude a fungal infection. Therefore, the most important function of the pan-fungal assay is identification of the pathogen, for instance when fungal elements are visible in clinical samples, but cultures remain negative. Furthermore, when a culture is positive, but the fungi cannot be identified by conventional methods, pan-fungal assays can be used for the

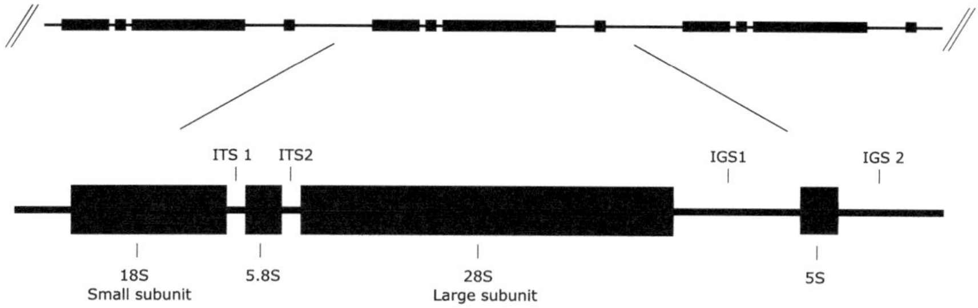

Fig. 4.1 Schematic illustration of the genes encoding the fungal ribosomal RNA, 18S, 5.8S, 28S and 5S. The ribosomal RNA genes (rDNA) of a fungal cell contains multiple copies of highly similar units of DNA, consisting 18S-ITS1-5.8S-ITS2-28 s-IGS1(-5S-IGS2). The locations of the internal transcript spacer (ITS) and intergenic spacer regions (IGS) are displayed. The 5S gene is lacking in most pathogenic moulds, those moulds contain only one IGS. The repeat unit contains highly conserved sites useful for the selection of universal primers for amplification and sequencing as well as regions with a high degree of variation useful for species and genus identification. The length of the region differs between fungal species; therefore the illustration's scale is only an approximation

molecular identification of the pathogen. The pan-fungal assays are often only available in specialized mycology reference centres.

Another approach is to use species- or genus-specific hydrolysis probes, targeting the variable regions within the rRNA genome with qPCR assays. This approach allows the use of the same universal primers, in combination with a specific probe for a targeted fungal species. Alternatively, the primers can be optimized for the targeted fungal species, allowing the use of smaller amplicons when compared to pan-fungal assays. Furthermore, the use of qPCR assays speeds up the detection, by avoiding the time consuming sequencing needed for sequence-based assays. A common disadvantage of the qPCR approach is that fungi, which are not targeted by the fungal specific hydrolysis, probe assays, will not be detected.

Thus, when a specific pathogen is suspected based on histopathological, clinical, or epidemiological grounds, a species- or genus-specific assay can be used. As the PCR assay can be optimized for the targeted species or genus, it often has a lower limit of detection than pan-fungal assays.

Next to the use of molecular methods for the diagnosis of invasive fungal disease, various qPCR assays have been evaluated for their use in monitoring progression of disease and responsiveness to therapy. Thus far, these assays have only been used in research settings, as the correlation between qPCR monitoring of fungal burden and the clinical outcome needs to be determined.

4.2.1.4 Choice of Specimen

Many opportunistic fungi enter the body by inhalation of fungal spores into the lungs. With an intact immune system, these spores are destroyed before they can invade human tissue. Most invasive fungal infections thus originate in the lungs. When fungal lung disease is suspected, a broncho-alveolar lavage (BAL) can be performed to obtain a sample from the infection site. Fluid is sprayed into suspect lung segments by bronchoscopy and again collected for analysis. Molecular detection of fungi in BAL samples has a high diagnostic sensitivity and specificity and specific PCR methods can be applied on the BAL samples based on microscopy and epidemiology. When BAL is precluded, due to the condition of the patient, sputum can be used for the isolation of fungal DNA, although sputum production is often absent in neutropenic patients. As a high concentration of fungal DNA is present in respiratory samples due to asymptomatic colonization, detection of fungal DNA by these qPCR assays does not necessarily imply that the patient suffers from a fungal infection and positive results should be interpreted in conjunction with clinical findings and other markers for invasive disease. For non-sterile sites, like sputum or BAL, the positive predictive value of PCR assays depends on the prevalence of fungal disease in the population and the type of specimen used. For example, isolation of *Aspergillus* DNA from sputum in an immunocompetent patient may resemble colonization, but in patients at high-risk for fungal disease, such as hematopoietic cell transplant recipients with clinical features of invasive disease, the positive predictive value for invasive fungal disease is much higher. Therefore, molecular methods should only be used when fungal disease is suspected on clinical grounds, as the interpretation of PCR results depends highly on the clinical features and radiological imaging findings. Other non-culture diagnostics like fungal antigens may help to increase the likelihood of the diagnosis as they detect another fungal target and are produced by growing hyphae. Galactomannan (GM, Aspergillus-antigen), is a component of the cell wall of *Aspergillus* species and can be detected in serum, BAL-fluid or cerebral spinal fluid. A positive GM in serum or BAL supports the diagnosis of invasive aspergillosis. β-D-Glucan is part of the cell wall of many fungal species, like *Aspergillus* and *Candida,* but not *Cryptococcus* or the Mucorales and is therefore a pan-fungal marker. Results should be interpreted carefully however; as the antigen is widely present potentially resulting in false positive reactivity.

When fungal elements are seen in normally sterile tissue obtained by biopsy, PCR can be used directly on tissue after DNA isolation for the identification of the causative species. A genus/species specific PCR can be used based on the microscopic morphology and local epidemiology. When the specific PCR remains negative, a pan-fungal assay can be applied to detect fungi not targeted by the specific PCR. Pan-fungal assay is less suitable to use on non-sterile respiratory specimen, as the assay will also detect potential contaminants from the upper airways.

Another promising approach is the use of PCR assays on whole blood, serum, or plasma. Whether whole blood, serum, or plasma is the best sample to detect fungal DNA in blood depends on the causative agent. More research is needed for recommendations on this matter.

4.2.1.5 Isolation of Fungal DNA

DNA extraction methods are presumably the limiting factor for molecular detection of fungi. Both moulds and yeast cells have very firm cell walls. Vigorous methods for cell wall lysis are needed and critical for optimal DNA isolation. To disrupt the cell walls adequately, methods, such as nucleotide extraction by phenol-chloroform or mechanical cell wall disruption by agitating the sample with beads, are needed prior to the automated methods used for DNA purification. The purification methods are similar to those used for bacteria and viruses and thus the same machines can be used.

For histopathology evaluation, obtained material is often formalin-fixed, and paraffin-embedded. Formalin induces crosslinking between DNA and proteins, hereby inhibiting the amplification of long DNA fragments. Several protocols have been developed for DNA isolation of formalin-fixed material, but amplification of DNA fragments of more than 200–300 base pairs is often impossible. PCR assays for formalin-fixed, paraffin-embedded material should ideally use short amplicons. Consequently, pan-fungal PCR assays, with long amplicons, have low performance on these samples.

4.2.2 Invasive Candidiasis

Candida species are commensal organisms of the human skin, mouth, and gastrointestinal tract. However, to immunocompromised patients, to those with critical illness, or to those that are subjected to abdominal surgery, *Candida* might cause invasive disease. Colonization is recognized as a risk factor for invasive disease, as infection is often preceded by yeast colonization. Invasive candidiasis involves both deep-seeded infections (the presence of candida species in deeper tissue) and candidemia, (presence of *Candida* species in blood).

The antifungal susceptibility differs among yeast found in invasive candidiasis. *Candida krusei* and a high proportion of *C. glabrata* isolates have reduced sus-

Table 4.1 Recommendations for the use of molecular methods for the diagnosis of invasive candidiasis

Guidelines on invasive candidiasis	ESCMID	IDSA	ECIL
Year of publication	2012	2016	2009
Recommendations regarding molecular detection of invasive candidiasis	No recommendations can be made due to lack of data	Candida PCR is of potential benefit, but its role in the management of invasive candidiasis is unclear. More data is needed	No recommendations were made for invasive candidiasis

ceptibility for fluconazole, whereas *C. parapsilosis* is less susceptible in vitro to the echinocandins. A 24-h delay of appropriate treatment for invasive candidiasis results in a significantly increased mortality, but the diagnosis is often missed as the diagnostic sensitivity of blood culture in candidemia is estimated to be only 50–75%. Therefore, a fast, sensitive, non-culture based method for the identification of causative agents in candidemia may be of clinical benefit. The use of PCR on blood samples has been shown to shorten the time to diagnosis and starting of antifungal therapy.

4.2.2.1 Molecular Detection of Invasive Candidiasis

Invasive candidiasis can be diagnosed by blood culture. In addition, surrogate markers such as antigen (mannan or β-D-glucan), antibodies (antimannan) and PCR are helpful in the diagnosis of invasive candidiasis. The mannan-antimannan testing has variable sensitivity. A higher diagnostic sensitivity is reported for PCR, which is similar to that of β-D-glucan. However, PCR has some advantages over β-D-glucan testing. A true positive β-D-glucan is not specific for C*andida* disease, while PCR enables both genus and species identification.

qPCR assays with multiple hybridization probes for the direct identification of several clinically important *Candida* species have been developed. Furthermore, pan-fungal assays can be used to identify *Candida* spp. The lower limit of detection (see Vol. 1, Chap. 6) of PCR assays used on blood is a very important determinant. The blood of more than half of patients with candidemia contains less than 1 *Candida* cells/ml. However, blood also contains free floating yeast DNA-fragments. In general, for optimal PCR performance, the lower limit of detection of the PCR method should at least be 10 CFU/ml. The best performing *Candida* PCR-assays are those that target the rRNA (Fig. 4.1) or the cytochrome P450 genes.

For molecular detection of candidemia, some studies indicate that whole blood is superior to serum. However, in a large clinical study, whole blood was shown to be inferior to serum and plasma, presumably because of the destruction of free floating DNA in the blood due to glass beating, which is needed for fungal DNA extraction from intact yeast cells. Thus, a final conclusion to define the best specimen is still to be drawn. The estimated diagnostic sensitivity and specificity of *Candida* PCRs on whole blood is around 85 and 90% depending on the tested population.

For deep-seeded infections, sterile tissue can be used for the direct detection of *Candida* DNA. A positive PCR might be used as proof for infection; the negative predictive value depends on the fungal load in the tissue.

Detection of Candida DNA in BAL, sputum or other non-sterile samples is not useful as a positive result may represent colonisation rather than invasive disease. Nonetheless, molecular methods might support the detection of yeast colonization in patients at risk for invasive candidiasis. As with most molecular methods for invasive fungal disease, data concerning the clinical application are generally lacking and PCR can only be used in addition to recommended methods for diagnosis.

For *Candida* species, molecular methods have been developed to detect secondary resistance markers. The mechanism of secondary echinocandin resistance is similar in most *Candida* species. It is due to target site mutations in the *FKS1* gene and *FKS2* gene. These mutations are strongly associated with echinocandin treatment failure. It is possible to rapidly detect target site mutations by real time PCR methods. Secondary azole resistance may be due to a variety of mechanisms, which hampers the use of molecular detection methods. Molecular methods have developed to detect target site mutations in the *ERG11* gene conferring azole resistance. Furthermore, overexpression of efflux pumps leading to azole resistance in *Candida* has been described. These mechanisms can be detected by measuring the level of transcript of the genes involved.

4.2.3 Invasive Aspergillosis

Aspergillus species are the most common isolated moulds in patients with opportunistic fungal infections. *A. fumigatus* is the most prevalent species in most geographic areas. However, in warmer regions, like India and Iran, *A. flavus* is the most isolated species. Other species causing human disease are *A. terreus*, *A. nidulans* and *A. niger*. Several antifungal agents are available for the treatment of invasive aspergillosis. Amphotericin B is a broad-spectrum antifungal agent, often advised as the first choice for empirical treatment of invasive fungal disease when the pathogen is not known or when zygomycosis is suspected. The first choice of treatment for invasive aspergillosis is voriconazole or isavuconazole. The *Aspergillus* species differ in the susceptibility to antifungal agents. Most *Aspergillus* species are susceptible to voriconazole, although in some areas, resistance to voriconazole is increasingly reported in *A. fumigatus*. Furthermore, several other

Table 4.2 Recommendations for the use of molecular methods for the diagnosis of invasive aspergillosis

Guidelines on invasive aspergillosis	IDSA	ECIL
Year of publication	2016	2009
Recommendations regarding molecular detection of invasive aspergillosis	PCR based diagnosis is promising, but the methods are not standardized or commercially available. Combining PCR and galactomannan or β-D-glucan is a important research direction. Results should be considered in conjunction with other diagnostic tests and the clinical context	The current status of the technical and clinical validation of PCR for Aspergillus in blood and other fluids does not currently allow for a recommendation for clinical use

Aspergillus species may be inherently less susceptible to various antifungal agents. For example, *A. terreus* has decreased in vitro susceptibility to amphotericin B. Although species identification cannot fully predict antifungal susceptibility, it may aid in clinical management and determining prognosis. When a fungal culture is available, most species can be identified by conventional methods only to the species-complex level. However, as with most fungal cultures, *Aspergillus* takes several days to produce the spore-forming units that are needed for species-complex identification. Furthermore, some cryptic species cannot be distinguished using conventional methods and may show distinct antifungal susceptibility profiles. For these reasons, species identification is important and may aid clinical decisions and in predicting prognosis.

4.2.3.1 Molecular Detection

Aspergillus can be identified using pan-fungal assays. In addition, several *Aspergillus* specific assays have been developed. Pan-*Aspergillus* kits, targeting multi-copy genes, have been developed for the identification of *Aspergillus* at the genus level. Post amplification sequence analysis of the amplified elements allows species identification. PCR can be applied on respiratory samples, like BAL or sputum. PCR on BAL has very high diagnostic sensitivity for the detection of *Aspergillus*. Meta-analysis shows that a negative PCR for *Aspergillus* species has a high negative predictive value for ruling out invasive aspergillosis. Therefore PCR on BAL samples can be used as a tool to exclude *Aspergillus* infection of the lung. However, studies on the performance of *Aspergillus* PCR assays are highly variable, due to variations in DNA isolation protocols and primer targets used, which is why the method is not widely incorporated in clinical practice. The diagnostic sensitivity and specificity of Aspergillus DNA detection in BAL are around 90 and 96%, respectively. Recent meta-analyses show that the combination of *Aspergillus*-specific PCR in combination with GM testing further increases the diagnositic sensitivity of invasive aspergillosis enabling early initiation of antifungal treatment.

In addition to the detection of *Aspergillus* species in respiratory samples, PCR can be performed on blood samples and tissue biopsy. PCR on plasma has shown to give a positive result more than 2 weeks before clinical diagnosis, outperforming PCR on serum. The diagnostic sensitivity en specificity of *Aspergillus* PCR on blood samples were 84 and 76%, respectively, in a meta-analysis. Furthermore, PCR performed on cerebrospinal fluid samples showed 100% diagnostic sensitivity for cerebral aspergillosis in a small study in six patients. Before molecular detection of *Aspergillus* species can be generally implemented in diagnostic criteria, research must aim to harmonize the methodology and to clinically validate *Aspergillus*-assays.

Molecular identification of cryptic *Aspergillus* species can be done by sequence analysis of the single-copy beta-tubulin or calmodulin gene. However, this method is only used for cultured isolates, due to low sensitivity for the single copy target.

Resistance mutations of *A. fumigatus* for medical triazoles (the first choice therapy for invasive aspergillosis) are well described and predominantly include substitutions in the *Cyp51A*-gene or its promoter region. Detection of resistance relies on phenotypic testing of *A. fumigatus* colonies, but due to the low diagnostic yield of culture resistant cases may be missed. Therefore, molecular methods have been developed to detect azole resistance mutations in culture or directly in clinical specimens. Sequence analysis of the full *Cyp51A*-gene and its promoter region can identify these mutations. Importantly, mutations correlated with azole resistance should be distinguished from polymorphisms. Furthermore, qPCR kits have been developed for direct detection of the TR_{34}/L98H and TR_{46}/Y121F/T289A mutations in clinical specimens. Furthermore, assays detecting *Cyp51A* substitutions at codon G54 and M220 have been described. Studies are ongoing that further evaluate the diagnostic performance of Aspergillus resistance PCRs.

4.2.4 *Mucorales,* Scedosporium, Fusarium *and Other Opportunistic Filamentous Moulds*

Discrimination between *Aspergillus* and other moulds, like *Scedosporium, Fusarium* and fungi belonging to the genera Mucorales, is important, because of different antifungal susceptibility patterns and prognosis. Mucorales belong to the Zygomycetes, a large order of primitive, fast growing fungi. Mucormycosis (or zygomycosis) primarily affect patients with immunodeficiency diseases and patients with dysregulated diabetes. Mucorales are resistant to voriconazole, so a lipid formulation of amphotericin B is recommended for initial antifungal therapy. However, the Mucorales may be difficult to culture from clinical specimens and due to the lack of specific biomarkers, diagnostic delay is common.

Several *Scedosporium* and *Fusarium* species are unresponsive to various antifungal agents, complicating the management of infections caused by these moulds. The most often isolated *Scedosporium* species, namely *S. prolificans* and *S. apiospermum,* show distinct antifungal susceptibility patterns, and may show different clinical manifestations as well as outcome. For example, *S. prolificans* (recently reclassified as *Lomentospora prolificans*) shows a more resistant antifungal profile than *S. apiospermum.* Furthermore, invasive infection accounted for about 50% of clinical *S. prolificans* isolations, while invasive infection accounted for less than 10% of *S. apiospermum* isolations in a large retrospective study from Australia. Scedosporiosis due to *S. prolificans* is often irresponsive to antifungal treatment alone and source control trough surgical intervention of often needed to control the infection. *Fusarium* is another important mould, which causes various manifestations of diseases in humans, including infection of the cornea. Dissemination is often rapid due to the fungus' ability to sporulate in tissue and reported mortality rates are high, 60–80%. Microscopic examination of BAL fluid, sputum, or tissue may show fungal mycelium. Specific stains, such as the optical

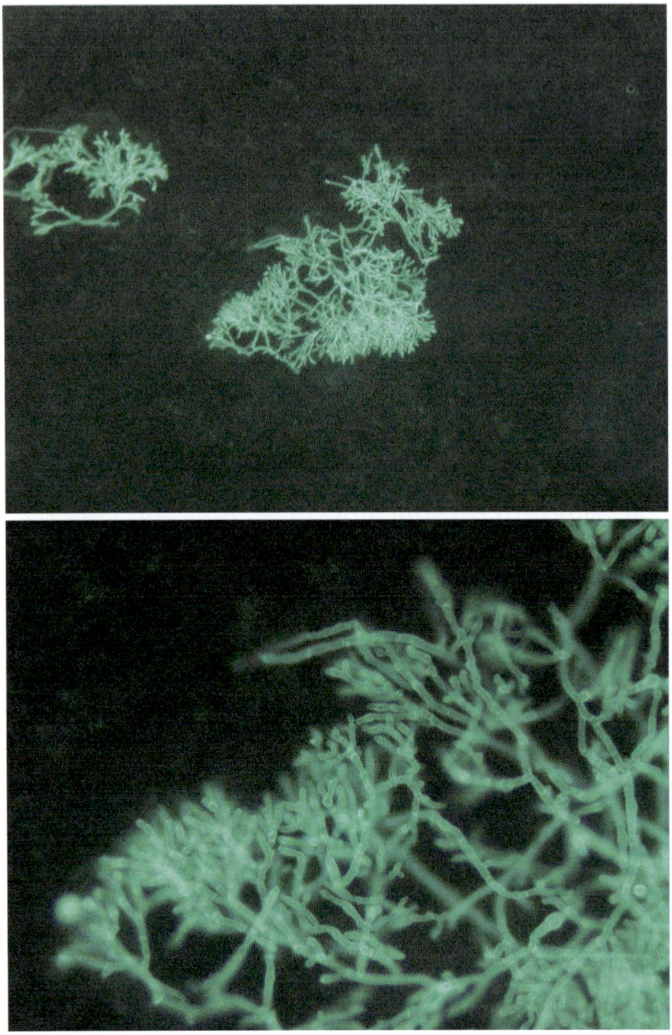

Fig. 4.2 Blankophor P fluorescence stain showing a septated mould in sputum of a immunocompromised patient (200 and 400x). The procedure can be used to detect fungal elements is various forms of clinical specimen. However, direct examination of clinical specimens does not allow identification of fungi to the species level

brightener Blankophor P, allow some discrimination of fungal pathogens. Non-septate hyphae are consistent with Mucorales and can be distinguished from septated moulds like *Aspergillus*. However, differentiation of *Scedosporium* or *Fusarium* from *Aspergillus* is not possible, as all three show similar microscopic characteristics in tissue (Fig. 4.2).

Table 4.3 Recommendations for the use of molecular methods for the diagnosis of invasive mucormycosis and hyalohyphomycosis

Guidelines on other opportunistic filamentous fungi	ESCMID	ECIL
Year of publication	*1 2013 *2 2014	*3 2013
Recommendations regarding molecular detection of other opportunistic filamentous fungi	*1 A standardized test is not available. Use of molecular methods is moderately supported *2 Molecular tests are helpful, but only to be used to supplement conventional laboratory tests	*3 No standardized test is available. No recommendation

*1 Guideline on mucormycosis; *2 Guideline on hyalohyphomycosis: *Fusarium* spp. and *Scedosporium* spp.; *3 Guideline on mucormycosis

4.2.4.1 Molecular Detection

Pan-fungal PCR assays targeting the genomic rRNA region are able to detect isolates of these genera, but species-specific methods have higher sensitivity. Rapid molecular detection enables timely adjustment of antifungal therapy if necessary, improving outcome significantly, especially if conventional methods fail to identify the fungal agent. Specific qPCR assays for the detection of Mucorales have been shown to be effective for screening of populations at risk for mucormycosis. However, as many Mucorales species are known to cause human disease and the prevalence differs greatly among regions, PCR assays should be customized based on local epidemiology. Genus-specific PCR methods for *Scedosporium* and *Fusarium* species have been developed, but are less useful in the clinical setting, as most invasive fungal infections are caused by *Aspergillus*.

Molecular identification alone cannot fully predict antifungal susceptibility or responsiveness to therapy and thus will not replace antifungal susceptibility testing. However, it may aid in early adaptation of clinical management and will give important information for the prediction of prognosis. Again, prospective clinical trials that have determined the diagnostic sensitivity and specificity of these PCR methods are lacking.

4.2.5 Pneumocystis jiroveci *Pneumonia*

Pneumocystis jiroveci is the most common opportunistic fungal infection affecting HIV-positive patients. Other patients with immunodeficiency are also at risk. Mortality of *P. jiroveci* pneumonia (PCP) is estimated at about 10–20% for HIV-positive patients and even higher for patients with other underlying diseases. The diagnosis relies on clinical symptoms, chest radiography, and detection of β-D-glucan in serum. For a definite diagnosis, determination of *Pneumocystis* spp. in

Table 4.4 Recommendations for the use of molecular methods for the diagnosis of pneumocystosis

Guidelines on *Pneumocystis jiroveci* pneumonia	ECIL
Year of publication	2016
Recommendations regarding molecular detection of *Pneumocystis jiroveci* pneumonia	Real-time qPCR is recommended on BAL samples in combination with immunofluorescence. PCP can be ruled out with negative PCR and immunofluorescence

respiratory samples following cytological staining or direct immunofluorescent staining is needed. *Pneumocystis* cannot be cultured, which presumably has enhanced the development of molecular detection methods. Although clinical validation is still incomplete, molecular methods for the detection of *Pneumocystis* have been around for about 20 years, being implemented in many microbiology laboratories nowadays. Studies indicate that PCR methods for the detection of PCP have better diagnostic sensitivity in comparison to staining methods, but the diagnostic specificity is generally lower. Therefore, the PCR often used as a screening method for PCP, followed by immunofluorescent staining to confirm the diagnosis.

4.2.5.1 Molecular Identification

The first molecular methods for the detection of PCP were nested PCR assays (see Vol. 1, Sect. 5.12.7), which could sensitively detect *Pneumocystis* DNA. Examples of genomic targets are the mitochondrial large subunit, the dihydropteroate synthase gene, and the heat shock protein 70 gene. Important drawbacks of the nested PCR include the inability to differentiate between colonization and infection, and the risk of contamination during the obligatory opening of the reaction tubes. The prevalence of asymptomatic carriage of *P. jiroveci* is estimated to be ~18%, thus the clinical utility of a nested PCR is relatively low and clear clinical features and radiographic imaging are needed for the diagnosis of PCP. qPCR methods have been developed both to detect *Pneumocystis* and to estimate the fungal load. However, cut-off values that differentiate between infection and colonization are difficult to establish. Various cut-off values are reported in literature. However, since genomic targets and PCR methodology differ greatly between clinical studies and centres, the published cut-off values cannot be extrapolated and preclude broad use in daily practice. Consequently, a definite and clinical useful cut-off for the differentiation of colonization and infection of *Pneumocystis* spp. is still to be validated.

4.3 Molecular diagnostics of dermatophyte infections

A.M.C. Bergmans

4.3.1 Introduction

The use of molecular techniques in the clinical microbiology laboratory is increasing almost exponentially, as these techniques are very fast, sensitive and specific when compared to classical methods such as microscopy and culture. Particularly in laboratory diagnostics in dermatophytosis, where detection and identification of the causative agent by culture methods is rather time-consuming (two to four weeks) and labor-intensive, the introduction of PCR has led to a major improvement with regard to speed. This results in faster, targeted treatment or in the avoidance of unnecessary antifungal treatment, preventing toxicity/adverse effects. Additionally, it results in improvement of infection control. In this chapter the major aspects of dermatophytes and their diagnosis are discussed, with an emphasis on molecular methods.

4.3.2 Dermatophytosis

Dermatophytes are closely related keratinophilic moulds, which can cause superficial infections involving the skin, hair and nails. Infections caused by dermatophytes are called tinea - followed by the site of infection. For example, tinea corporis, involving the arms, trunk and legs, tinea capitis, involving the scalp, and tinea pedis, involving the foot. Fungal nail infection, called onychomycosis, is the most common nail disorder in adults worldwide. Based on their ecology, dermatophytes have been divided into three groups: anthropophilic, zoophilic and geophilic species. Anthropophilic species also occasionally infect animals, and zoophilic species occasionally infect humans. Geophilic dermatophytes are associated with keratinous materials (hair, feathers, horns) present in soil.

4.3.3 Taxonomy and Epidemiology of Dermatophytes

Three genera are known among the dermatophytes: *Trichophyton*, *Microsporum*, and *Epidermophyton*. The taxonomy of dermatophytes has undergone significant changes during the last 15 years, as a consequence of the characterization of the dermatophyte genome. The gold standard for dermatophyte species identification is now based on sequencing of the internal transcribed spacers (ITS) of the ribosomal rRNA operon. Using these sequences, dermatophytes were grouped based on their clinical and ecological traits, rather than their morphological features. The most common anthropophilic species include *Trichophyton rubrum*, *Trichophyton interdigitale*, *Trichophyton tonsurans*, *Trichophyton violaceum*, *Trichophyton soudanense*, *Trichophyton concentricum*, *Trichophyton schoenleinii*, *Microsporum audouinii* and *Epidermophyton floccosum*. Zoophilic species (that occasionally infect humans) include *Microsporum canis* (cats, dogs, horses), *Trichophyton mentagrophytes* (rodents, dogs, horses, mice, camels), *Trichophyton verrucosum* (cattle), *Trichophyton erinacei* (hedgehogs, rabbits, guinea pigs, other rodents) and *Trichophyton simii* (monkeys). *M. audouinii*, *T. tonsurans* and *T. soudanense* are known to cause mainly tinea capitis. Other species including *T. rubrum* and *T. interdigitale* are found mostly in onychomycosis. *E. floccosum* only infects the skin. The distribution of dermatophytes varies considerably, depending on geographical area, climate and other epidemiological factors (i.e. age, sex, lifestyle, seasons). In Western countries, tinea pedis and onychomycosis are the most prevalent clinical forms of dermatophytosis, while in tropical regions, tinea capitis and tinea corporis are more prevalent. The proportion of different dermatophyte species in clinical samples as determined both by real-time PCR and by culture in more than 6,200 clinical samples between 2007 and 2011 in The Netherlands is shown in Table 4.5. The predominance of *T. rubrum* suggests a great capacity of this species to spread. Recent molecular studies have shown that this is the case for at least some strains of *T. rubrum*. The most common source of infection with *T. rubrum* is the bath tub that is contaminated by family members. However, other factors, like the increased use of sports facilities, including public pools, changing rooms, showers and mats, and the use of occlusive footwear which causes humidity, may explain the rise of this species. The introduction of griseofulvin in 1958 as a systemic antifungal agent

Table 4.5 Subdivision of different dermatophyte species in Dutch clinical samples as determined in positive clinical samples by real-time PCR ($n \approx 2850$) and by culture ($n \approx 1650$) between 2007 and 2011

	PCR (n)	Culture (n)	PCR (%)	Culture (%)
T. rubrum	2172	1226	75.6	74.3
T. interdigitale	605	326	21.1	19.8
T. tonsurans	41	40	1.4	2.4
M. canis	14	10	0.5	0.6
T. erinacei 2	12	12	0.4	0.7
M. audouinii	9	8	0.3	0.5
E. floccosum	6	4	0.2	0.2
T. violaceum	5	10	0.2	0.6
T. verrucosum	1	1	0.03	0.1
T. species	3	2	0.1	0.1
T. rubrum + T. interdigitale	4	10	0.1	0.6
T. interdigitale + T. violaceum	1	0	0.03	0.0
T. mentagrophytes	0	1	0.0	0.1
Total	2873	1650	100	100

for the treatment of tinea corporis and tinea capitis resulted in the disappearance of *M. audouinii* and *T. schoenleinii* from central Europe. *T. rubrum* is the most frequently detected species worldwide, while *M. canis* is still the main causative agent in tinea corporis and tinea capitis in Mediterranean countries. The prevalence of anthropophilic dermatophytes causing tinea capitis in young children is increasing in the big cities of Europe. This increase has occurred in parallel to the increase in immigration from countries with endemic tinea capitis.

4.3.4 Conventional Laboratory Diagnosis of Dermatophytosis

Conventional laboratory diagnosis of infections caused by dermatophytes relies on microscopic examination of KOH-treated clinical samples (with optional addition of calcofluor white), microscopic and macroscopic observation of in vitro cultures, and metabolism tests. Microscopic examination is nonspecific as dermatophytes as well as non-dermatophytes will be detected. In addition, with microscopic examination, no identification of the fungus is possible.

Culturing on selective media has low sensitivity (50–70%), is time-consuming due to the slow growth rate of dermatophytes (two to four weeks), and identification of cultured isolates can be difficult. Different morphological characteristics and structures, such as colony pigmentation, texture, growth rate, presence of microconidia and/or macroconidia, spirals, pectinate branches and other structures, are used for identification. Therefore, detection and identification of dermatophytes based on morphology and colony characteristics of cultured isolates is very labor-intensive and it requires a high level of scientific knowledge and experience.

4.3.5 Molecular Methods for Detection of Dermatophytes in Clinical Material

4.3.5.1 DNA Extraction from Clinical Material

The most important part in the procedure of DNA extraction from clinical dermatophytosis samples is the efficient lysis of the tough, keratin-containing nail, skin and hair material that is used for molecular diagnostics. Additionally, the procedure needs to be able to to disrupt the rigid chitin containing fungal cell wall. Another hurdle is the removal of possible PCR inhibitors present in this kind of clinical material. Until recently, various approaches have been used for homogenization and lysis, such as manual or freeze/thaw processing, bead-beating or phenol-chloroform extraction.

More recently, overnight proteinase K digestion (with or without dithiothreitol as a detergent of keratin) followed by commercial manual or automated kit extraction is increasingly being used in routine clinical laboratories. In addition, a 15-min, two-step buffer method has been described recently for rapid DNA extraction from clinical material.

The use of an internal process control (IPC; see Vol. 1, Chap. 6) during the whole extraction and amplification process is a must when performing molecular diagnostic assays. It generally consists of a DNA or RNA target (for example a virus) that is added to every clinical sample before DNA extraction, and is co-amplified with the target in question to detect false-negative assay results (see Vol. 1, Chap. 6).

4.3.5.2 Targets for Molecular Detection of Dermatophytes in Clinical Material

The chitin synthase 1 (CHS1) gene is highly conserved among dermatophyte species, so it has been the target for pan-dermatophyte molecular assays for several years. However, due to its low variability, this gene is not useful for species identification. Only a very small number of assays have been described, which use the DNA topo-isomerase II gene or a microsatellite marker for the detection of dermatophytes. Currently, almost all molecular assays for species-specific detection of dermatophytes use targets located in the ribosomal RNA operon (covering 18S, ITS1, 5.8S, ITS2 and 26S or LSU). This operon is selected because of the presence of highly variable regions in multiple copies, greatly improving the analytical sensitivity of the diagnostic assay.

4.3.5.3 Amplification and Detection Methods

In the last decennium, several diagnostic methods for the direct detection and identification of dermatophytes from clinical samples using molecular biology-based methods have been published. Many different techniques have been used. Among them is (multiplex) genus- or species-specific PCR, where species identification is achieved by evaluating the PCR product lengths (with or without digestion by restriction enzymes) visualized on an agarose-gel. Another technique is PCR-reverse line blot, where PCR products of conventional PCRs are analyzed by hybridization with dermatophyte genus- and species-specific immobilized probes. The subsequent detection is achieved by streptavidin-peroxidase and chemiluminescence on a light-sensitive film. Others have described nested PCR, sometimes followed by sequencing of the product for identification of the detected species. PCR-ELISA has also been described, which is a PCR amplification followed by hybridization of digoxigenin-labeled PCR products to an array of biotin-labeled probes immobilized in 96-well plates. These techniques all are very laborious, time-consuming and prone to cross-contamination due to the need for post-PCR analysis.

In the last few years, more real-time PCR assays have been developed. Some of them use sets of species-specific primers and probes, others consist of multiplex PCRs with pan-dermatophyte primers, combined with species-specific primers and/or probes for *T. rubrum* (and some with extra probes for *Trichophyton* spp. and *M. canis/audouinii,* or *T. interdigitale*, *T. tonsurans*, *T. violaceum* and *Microsporum* spp.) or pan-dermatophyte nested PCR. Almost all published assays use hydrolysis probes for the detection and identification of the PCR products.

One PCR assay will be described here in more detail, as it is the most comprehensive real-time PCR assay, which allows for the simultaneous detection and discrimination of 11 common, clinically relevant, dermatophyte species in a single tube. In contrast to most other real-time PCR assays described, this assay uses melting curve analysis with hybridization probe sets, instead of hydrolysis probes for the identification of species.

Clinical samples, including nail, skin and hair, are cut into small pieces and incubated overnight at 55 °C in a lysis buffer containing dithiothreitol and proteinase K (in order to completely lyse the tough, keratin-containing nail, skin and hair samples). After addition of phocine herpes virus 1 (PhHV-1) DNA as an IPC, total nucleic acid is isolated on a MagNA Pure (Compact) isolation robot. Dermatophyte-specific PCR primers and species-specific LightCycler hybridization probe sets for real-time PCR (each consisting of an anchor probe and a sensor probe, see also Vol. 1, Chaps. 4 and 5) target dermatophyte ITS1 and 5.8S rRNA gene sequences. The probes used in this assay are labeled with 3'-fluorescein, 5'-LC610, 5'-LC640, 5'-LC670, or 5'-LC705 labels. Anchor probes are designed to perfectly match ITS1 sequences of two or three dermatophyte species, whereas sensor probes have a perfect match with only one of those species, and have 1-3 mismatches with the other species that match with the anchor probe (Fig. 4.3). With this approach, up to three different dermatophyte species can be detected and identified with one hybridization probe set, using melting curve analysis (see also Vol. 1, Chap. 5). Additionally, IPC (PhHV-1) primers and probes developed by Niesters are adapted to hybridization probe format. Real-time amplification and subsequent melting curve analysis is performed with two primer sets and six hybridization probe sets in a single tube. The presence of an amplification curve in the respective channels (610, 640, 670 and 705 nm) of the LightCycler 2.0, in conjunction with a melting curve with the appropriate melting temperature (T_m), is considered to be a positive PCR result for a given dermatophyte species (Fig. 4.4)

```
T. interdigitale   CGAGCCTCTC TTTAGTGGCT AAACGCTGGA CCGCGCCCGC CGGAGGACAG ACGCAAAAAA 260
T. mentagrophytes  .......... .......... C......... .......... .......... ..........
T. tonsurans       .......... .A...C.... C......... .......... .......... ..........
```

Fig. 4.3 Multiple sequence alignment of partial internal transcribed sequence I regions from three of the 11 dermatophyte species detected and identified by the LightCycler real-time PCR assay. Nucleotides corresponding to the species-specific hybridization probe sets are bold and underlined

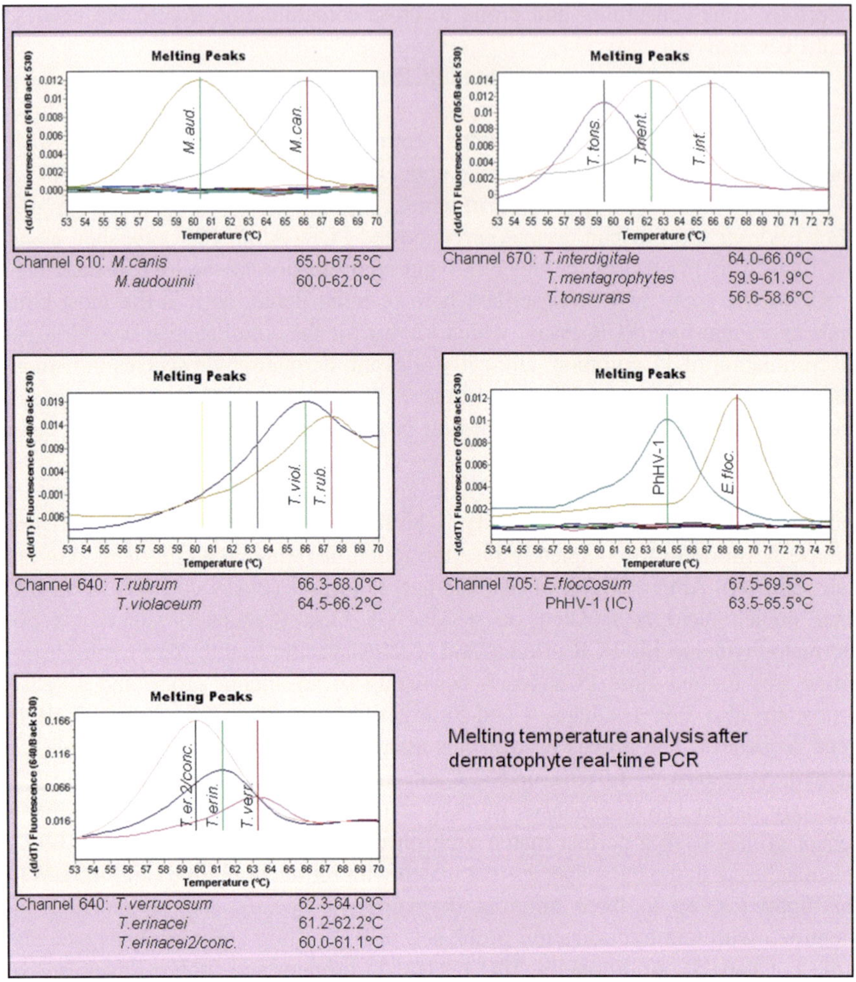

Fig. 4.4 Identification of 11 dermatophyte species and PhHV (internal control) by melting curve analysis with hybridization probe sets after real-time PCR amplification on DNA isolated from nail, skin and hair samples

[see suggested reading Bergmans et al., CMI 2010]. It needs to be mentioned that T_m ranges can vary depending on the extraction method, PCR master mix, real-time PCR instrument type, etc. Therefore, each laboratory should determine its own T_m ranges for each species under its own conditions.

4.3.6 *Suggested Algorithm for Diagnostics of Dermatophytes*

Based on the results of microscopy, culture and real-time PCR in a very comprehensive set of more than 5800 clinical samples (nail, skin and hair) collected during a 5-years period from 2007 until 2011 in The Netherlands, a new algorithm for laboratory diagnosis of dermatophytosis can be proposed (Fig. 4.5). One should start with real-time PCR, which results in dermatophyte-positive results in approximately 46% of all samples that can be reported directly as a definitive positive result to the clinician. The remaining 54% of all samples (i.e. PCR negatives) should be examined microscopically. All microscopy-negative (and PCR-negative) samples can be definitely reported as dermatophyte-negative to the clinician (48% of all samples). The samples that are positive by microscopy (and PCR-negative, 6% of the total number of samples) should be subjected to fungal culture. About 1% of all samples will yield dermatophyte culture-positives in this subset of PCR-negative, microscopy-positive samples. In approximately 12% of all samples, other, non-dermatophyte moulds and/or yeasts will be detected by culture (mono-infection or co-infection with a dermatophyte). Using this algorithm about 1% of all analyzed samples will be false-negative for dermatophytes (samples that were PCR-negative, microscopy-negative and culture-positive). If that remaining 1% needs to be detected, one should analyze all PCR-negative samples by culture (54% of total, compared to 6% of all samples if only PCR-negative, microscopy-positive samples are cultured) which will result in a substantial increase in costs and labor for dermatophyte diagnostics.

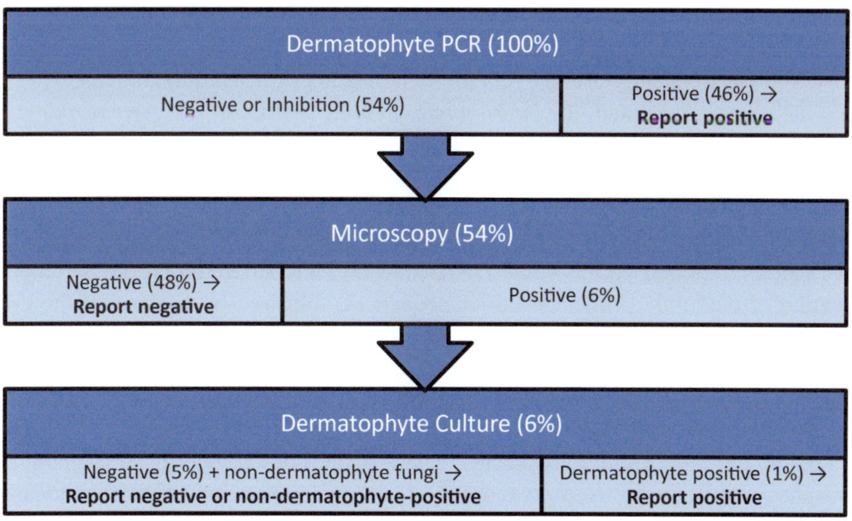

Fig. 4.5 Suggested algorithm for diagnostics of dermatophytosis, based on the results of a comprehensive comparative study on more than 5,800 clinical materials using dermatophyte culture, microscopy and real-time PCR in The Netherlands, 2007-2011 (unpublished data)

4.3.7 Quality Control

Methods for molecular diagnostics need to be very sensitive in order to detect even very low amounts of dermatophyte DNA. Another important requirement for molecular diagnostic assays is a high specificity, as it is unacceptable to have cross-reactions with other related species or with human DNA that is present in the clinical material to be analyzed. Therefore, thorough sensitivity and specificity testing of primers and probes should be performed in order to prevent false-positive and false-negative assay results. Additionally, optimal DNA extraction and the use of an internal control should also reduce the number of false-negative results. As PCR-based diagnostic assays are prone to cross-contamination of samples by PCR products from previous amplifications, the use of three separate rooms for preparation and storage of PCR reagents (1), DNA extraction and storage (2), and PCR amplification/detection (3) are essential to reduce the possibility of false-positive results. All aspects mentioned above can contribute to a sensitive and specific molecular detection assay. In 2016, the QCMD has distributed a new pilot proficiency panel for quality control of molecular diagnostics of dermatophytosis. This proficiency test panel will also contribute to the quality improvement of molecular assays for dermatophyte diagnostics.

4.3.8 Molecular Methods for the Identification of Dermatophyte Isolates

Many of the methodologies described for the identification of dermatophyte isolates are based on PCR and sequence analysis of the ITS region of dermatophytes, as this is the most accurate and discriminatory way to identify all clinically relevant species. Less frequently used targets for identification of isolates include 28S rRNA (*T. mentagrophytes* and *T. interdigitale*), chitin synthase (*T. mentagrophytes* complex, *T. rubrum*) and DNA topoisomerase (*T. rubrum*, *T. mentagrophytes*, *T. tonsurans*, *M. canis*, *M. gypseum*, *E. floccosum*). As sequence analysis is rather labor-intensive and expensive, other methods for the identification of isolates have emerged during the last two decades. Arbitrarily primed PCR (AP-PCR), rep-PCR or random amplified polymorphic DNA (RAPD) PCR all are based on the amplification of randomly primed repetitive sequences, followed by analysis of the different patterns of fragments obtained in different species. Another pattern-based identification method is PCR-RFLP, using DNA topoisomerase II, the hypervariable V4 domain of 18S rRNA + ITS, or the ITS region as a target. In PCR-RFLP, the PCR products are digested by restriction enzymes, which results in different fragment patterns in different species. These pattern-based identification methods are cheap and can be used successfully for the identification of most clinically relevant dermatophyte species. Another suitable method for the identification of isolates is PCR amplification of ITS1 and ITS2, followed by hybridization of the digoxigenin-labeled PCR products to an array of (genus- and) species-specific oligonucleotides on a nylon membrane. With this method, many isolates can be analyzed at the same time but it has a major disadvantage, as it is rather labor-intensive. Some ITS-based diagnostic real-time PCR assays that can discriminate between many dermatophyte species can also be used for the identification of cultured isolates.

Limited work has been done on intraspecies variability testing of isolates. Using RAPD PCR, intraspecies variability can be distinguished for *T. rubrum*, some *T. mentagrophytes* varieties, *E. floccosum*, *M. canis* and *M. audouinii*, but not for *T. tonsurans*. PCR-RFLP, based on 18S rRNA and ITS regions, can recognize several different patterns among isolates of *T. rubrum* and *T. interdigitale*. Recently, a PCR melting profile method, using restriction fragments, has been described for intraspecies differentiation of *T. rubrum* and *T. interdigitale*.

4.3.9 Conclusion and Final Remarks

Due to the application of molecular methods in exploring the taxonomy of dermatophytes, enormous progress has been made in uniform, objective species identification. As a consequence, many PCR assays for the direct detection and identification of dermatophytes in clinical material have been developed during the last decade. The introduction of real-time PCR assays in routine clinical laboratories has greatly reduced the turnaround time. Another major advantage real-time PCR of is the superior sensitivity as compared to dermatophyte culture. The development of molecular diagnostics has greatly improved the possibilities for faster, more guided treatment of dermatologic disorders, resulting in a significant decrease of the use of antifungals.

Recently, whole genome sequencing of *T. rubrum*, *T. tonsurans*, *T. equinum*, *M. canis* and *M. gypseum* has indicated features that differentiate these dermatophytes from other fungi and from each other. The dermatophytesseem to possess 308 genes that are completely unique. Additionally, each species has a variable number of species-specific genes, some of which play a role in determining host specificity. These genes may be perfect candidates for the development of new species-specific primers and probes for diagnostic purposes. The whole genome analysis also revealed gene families that are involved in mechanisms used by dermatophytes in causing chronic infections, in interactions with epithelial cells and with the host immune response. Some domains were overrepresented in dermatophytes. These domains are all involved in fungal growth on keratinized tissues, adaptation to skin, toxins or signals in the interaction between fungus and host, and avoidance of the host's immune response. These new developments may provide new insight into the factors that lead to fungal skin infection and their successful resolution following proper antifungal treatment.

Suggested Reading Invasive Fungal Infections

Alanio A, et al. ECIL guidelines for the diagnosis of Pneumocystis jirovecii pneumonia in patients with haematological malignancies and stem cell transplant recipients. J Antimicrob Chemother. 2016;71(9):2386–96.

Avni T, Leibovici L, Paul M. PCR diagnosis of invasive candidiasis: systematic review and meta-analysis. J Clin Microbiol. 2011;49(2):665–70.

De Pauw B, et al. Revised definitions of invasive fungal disease from the European Organization for Research and Treatment of Cancer/Invasive Fungal Infections Cooperative Group and the National Institute of Allergy and Infectious Diseases Mycoses Study Group (EORTC/MSG) Consensus Group. Clin Infect Dis. 2008;46(12):1813–21.

Halliday CL, et al. Molecular diagnostic methods for invasive fungal disease: the horizon draws nearer? Pathology. 2015;47(3):257–69.

Kourkoumpetis TK, et al. Polymerase chain reaction-based assays for the diagnosis of invasive fungal infections. Clin Infect Dis. 2012;54(9):1322–31.

McCarthy MW, Walsh TJ. PCR methodology and applications for the detection of human fungal pathogens. Expert Rev Mol Diagn. 2016;16(9):1025–36.

Mikulska M, Furfaro E, Viscoli C. Non-cultural methods for the diagnosis of invasive fungal disease. Expert Rev Anti Infect Ther. 2015;13(1):103–17.

White PL, et al. Aspergillus PCR: one step closer to standardization. J Clin Microbiol. 2010;48 (4):1231–40.

White PL, et al. Aspergillus polymerase chain reaction: systematic review of evidence for clinical use in comparison with antigen testing. Clin Infect Dis. 2015;61(8):1293–303.

Suggested Reading Dermatophyte Infections

Bergmans AMC, et al. Validation of PCR-reverse line blot, a method for rapid detection and identification of nine dermatophyte species in nail, skin and hair samples. Clin Microbiol Infect. 2008;14:778–88.

Bergmans AMC, et al. Evaluation of a single-tube real-time PCR for detection and identification of 11 dermatophyte species in clinical material. Clin Microbiol Infect. 2010;16:704–10.

Cafarchia C. et al. Molecular epidemiology, phylogeny and evolution of dermatophytes. Infect Genet Evol. 2013; 20:336–51 (Review).

Hayette M.P. et al. Dermatophytosis, trends in epidemiology and diagnostic approach. Curr Fungal Infect Rep. 2015; 9:164–79 (Review).

Jensen R.H. et al. Molecular diagnosis of dermatophyte infections. Curr Opin Infect Dis. 2012; 25:126–34 (Review).

Chapter 5
Molecular Diagnostics in Pathology

J.W.J. Hinrichs, R.A. de Weger, W.T.M. van Blokland, G.J.A. Offerhaus and P.J. van Diest

Abstract Molecular diagnostics is a rapidly growing discipline within pathology. This is largely explained by the changes in cancer treatment in the last decade. More and more types of cancer can be linked to specific DNA abnormalities and qualify for so-called targeted therapy. New techniques, such as next generation sequencing, allow tumour characterizations on the basis of DNA and/or RNA profiles. Moreover, the high sensitivity of digital PCR techniques enables DNA and RNA analysis on bodily fluids: the so called liquid biopsies. These developments will alter pathology practice dramatically in the coming years.

Keywords Pathology · Tumour · Somatic mutation · Targeted therapy · Next generation sequencing

5.1 Introduction

In pathology, molecular biological techniques are used to demonstrate qualitative and quantitative changes in human DNA or RNA, as well as for the detection of DNA or RNA of microorganisms in pathologically altered tissues or cells. Molecular pathology is mainly used for the optimization of tumour classification, but it is also becoming increasingly important in predicting effectiveness of specific (expensive anti-cancer) therapies. Successful molecular diagnostics in pathology depends on a close collaboration between pathologists and molecular biologists.

This chapter gives an overview of the different target areas of molecular diagnostics within pathology.

Because the detection of germline (inherited) changes belongs to the field of clinical and molecular geneticists, pathology is limited to the determination of acquired genetic (somatic) changes. Pathological genetic diagnoses often initiate

J.W.J. Hinrichs (✉) · R.A. de Weger · W.T.M. van Blokland · G.J.A. Offerhaus
P.J. van Diest
UMCU Utrecht, Utrecht, The Netherlands
e-mail: j.w.j.hinrichs@umcutrecht.nl

E. van Pelt-Verkuil et al. (eds.), *Molecular Diagnostics*,
DOI 10.1007/978-981-10-4511-0_5

germline mutational research and therefore has an important signalling role. Somatic DNA changes do not only play an important role in the development of cancer, but also form the basis of other defects.

Nowadays, many different techniques are used in molecular pathology, such as polymerase chain reaction (PCR) based techniques, in situ hybridization, DNA (cDNA) sequencing and a variety of DNA and RNA expression (Micro) array methods. Especially the PCR-based techniques have taken off in recent years. Reverse transcriptase (RT)-PCR, real-time quantitative PCR (qPCR), high resolution melting PCR analysis (HRMA) and various forms of multiplex-PCR are established applications in molecular pathology. Next generation sequencing (NGS) is a relatively new technique, with great promise for the future. The extremely high sensitivity of the PCR (wherein the sensitivity is not so much limited by the technique, but rather by the number of available target cells) allows for the detection of defects in a very small number of cells, or even in the DNA in blood, bone marrow, urine, saliva, or faeces, as well as cells isolated from tissue sections with the aid of laser-microdissection. This, however, increases the risk of false positive reactions, for example due to contamination or aberrant expression in benign cells. Morphological control of the analysed material by the pathologist therefore remains important.

5.2 Specific Molecular Pathology Conditions

Molecular diagnostic analyses in pathology are mostly performed on DNA and/or RNA that has been isolated from tissues. These tissues may be fresh or frozen, but are mostly fixed (usually in formalin) and embedded in paraffin. To ensure the quality of DNA and RNA, freezing or fixation of the tissue needs to happen as quickly as possible. Especially in tissues obtained after autopsy of a deceased person, as autolysis in tissue can result in insufficient DNA or RNA quality for further analysis. Freshly frozen tissue provides the best quality of DNA and RNA and this starting material is therefore suitable for virtually all techniques. Cytological preparations (smears and cytospin preparations) usually deliver excellent DNA. Fixed tissue does have limitations. In particular, fixation of the tissue in acidic fixatives (e.g. non-buffered formalin) is often disastrous for the quality of the DNA. Fixation and embedding of tissue results in fragmentation of the DNA, which makes it no longer possible to use techniques that depend on large DNA fragments. The degree of fragmentation can be determined using a multiplex PCR that amplifies specific fragments of increasing size (Fig. 5.1). On the basis of this analysis, it can be determined which techniques are still applicable. If the DNA is highly fragmented, the PCR strategy has to be adjusted to produce only smaller PCR products. If possible, techniques are used that work relatively well on fragmented DNA. One example of such a technique is MLPA, which makes use of hybridization probes for the detection of DNA alterations (see Vol. 1, Chap. 4) (Fig. 5.9).

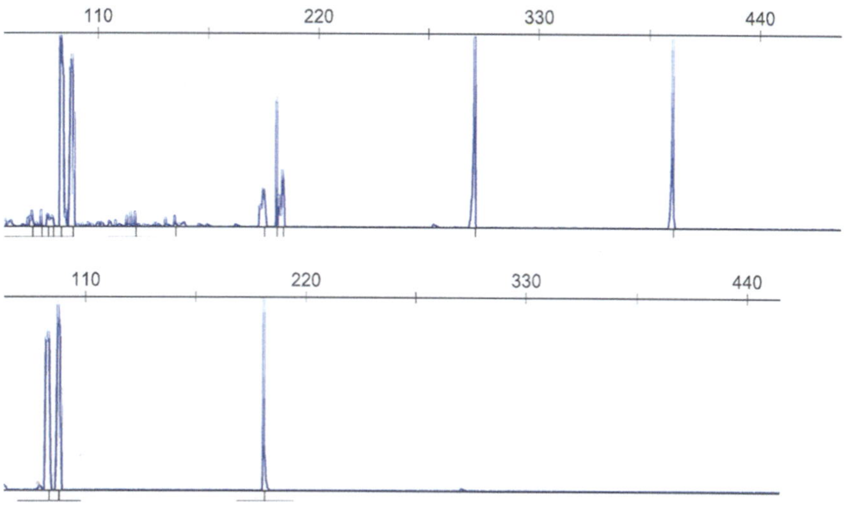

Fig. 5.1 DNA-fragmentation. The degree of DNA fragmentation can be determined with a multiplex-PCR amplifying specific fragments of increasing size. The used PCR primers are directed against general genes that are not associated with disease, so-called housekeeping genes. Because the forward primers are fluorescently labelled, fragment analysis can be done with a sequencer. In the obtained genescan, the fluorescence is plotted against the fragment size. In this example, the multiplex-PCR is performed on DNA isolated from two different formalin-fixed paraffin sections. In the upper genescan, PCR products of up to 400 basc pairs may be amplified, which is acceptable for formalin-fixed tissue. In the lower genescan PCR products up to 200 base pairs are the highest achievable, which indicates a high level of fragmentation of the DNA. This is possibly caused by excessive fixation with formalin

Isolation of DNA from paraffin-embedded material is usually carried out by means of a proteinase K lysis of paraffin sections, mostly followed by DNA purification with commercial columns. RNA can be obtained in a comparable way, although the quality of this purified RNA is often not optimal.

While DNA in situ hybridization can be performed equally well on both paraffin and frozen sections, RNA in situ hybridization is preferably carried out on frozen sections.

5.3 Molecular Diagnostics on DNA Without Genetic Changes

As will be seen in the following paragraphs, most pathology diagnostics is focused on genetic changes, especially in tumorigenesis. However, there are two applications in which non-genetically altered DNA is detected, namely in tissue identification and in the detection of viral/bacterial DNA.

5.3.1 Molecular Diagnostics for Tissue Identification

An important application in pathology diagnostics is the identification of possible mixed up tissues or so called "floaters" (contamination with foreign tissue in a slide or tissue cassette). The best way to do this is by PCR-analysis for informational (Person-specific), polymorphic DNA repeats (microsatellites or short tandem repeats (STRs)) (see Vol. 1, Chap. 9). DNA of both the suspect tissue and the initial patient tissue is collected from the slide(s). Then, a multiplex-PCR reaction is performed on multiple STR's, located on different chromosomes. The specific lengths of the STR PCR products are determined by means of fragment analysis. The array of STRs of the original and the suspect material will result in two unique DNA-fingerprints (Fig. 5.2). Comparing the two fingerprints will determine whether the suspect material matches that of the patient or if it has a different origin.

5.3.2 Molecular Diagnosis of Microorganisms in Pathology

Still, many diseases are caused by microorganisms. The detection of microorganisms in blood, urine, and faeces, for example, belongs to the field of medical microbiology. Conventional culture methods, often in combination with mass spectrometry (see Vol. 1, Chap. 1), still play an important role. However, in many cases, only histological or cytological material is available from a patient, which is primarily offered to the pathologist. It also happens that amount of the available patient material is insufficient for both morphological and conventional microbiological analysis. This material can then be tested for the presence of microorganisms in a pathology diagnostics laboratory. Various techniques are available, like histochemical staining (e.g. Ziehl Neelsen for *Mycobacterium tuberculosis*), immune histochemical/cytochemical techniques (for *Helicobacter pylori*) and electron microscopy (for viruses). Unfortunately, these methods are not always sensitive or specific enough.

Through the introduction of molecular biological techniques, such as in situ hybridization and PCR, the detection of microorganisms has changed dramatically in recent years. These techniques enable specific and sensitive detection of microorganisms in tissues and cytological material. Furthermore, these techniques have revolutionized the detection of poor or non-culturable microorganisms, such as the human papilloma virus (HPV). In situ hybridization has the advantage of being a morphological technique, and reaction products indicative of the pathogens are visible in infected cells or tissues. However, it has limited analytical sensitivity. A positive result can only be expected when the target is present in high copy numbers.

Molecular diagnostics of microorganisms within pathology, primarily relates to the detection of HPV, Epstein Barr virus (EBV), human herpes virus 8 (HHV8), Cytomegalo-virus (CMV), herpes simplex virus (HSV), *Chlamydia* species,

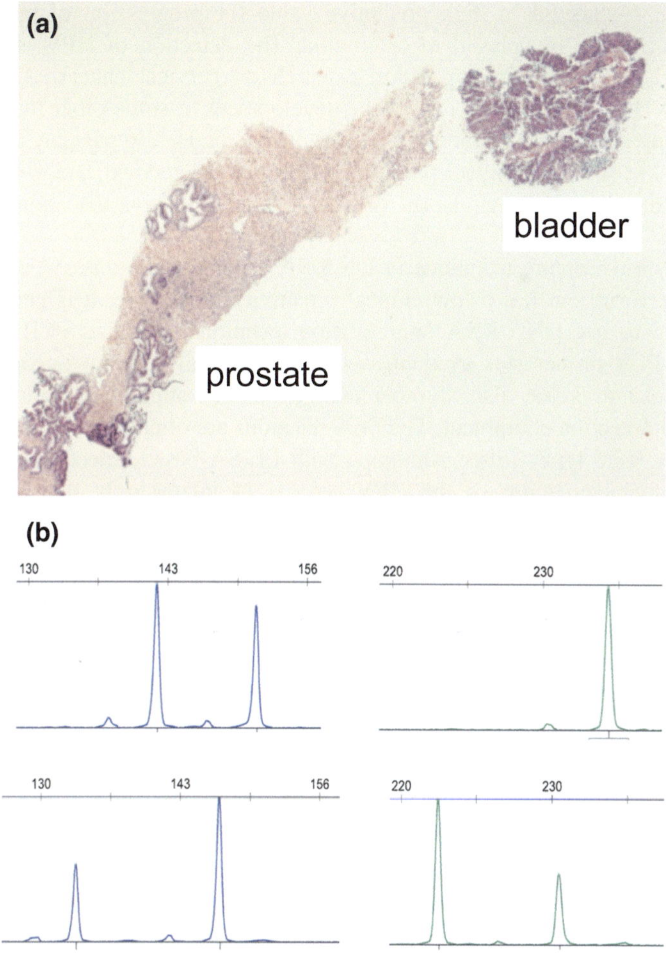

Fig. 5.2 Problems of mixed up tissues. **a** Tissue slide of a prostate biopsy with a "floater" of the bladder. **b** Tissue identification is carried out with a kit in which 16 STRs are analysed of which 2 STRs are shown. The STRs of the prostate material (*above*) are not identical to those of the bladder material (*below*) and thus not derived from the same person

Borrelia burgdorferi, *Mycobacterium* species, *Tropheryma whipplii* and *Bartonella henselae*. These tests can help to differentiate between certain types of tumours, or to identify specific infections.

HPV

HPV can inactivate the important tumour suppressor genes *Rb* and *p53*, which results in an increased life span of cervical epithelial cells. The presence of an oncogenic HPV type (e.g. types 16, 18, 31, 33) in the epithelium of the cervix is a prerequisite for getting cervical cancer. The persistent presence of oncogenic HPV in cervical smears has a high predictive value for progression of premalignant cervix aberrations (dysplasia) to carcinoma. The detection of HPV is becoming increasingly important in population research on cervical cancer. Additionally, oncogenic HPV types play a role in the development of some other malignancies, such as lung and tonsil carcinoma. The non-oncogenic HPV types play a role in the onset of skin warts, laryngeal papillomas and benign cervical and vulvar abnormalities. Since the HPV viral load is often small, a sensitive detection method is important.

The Hybrid Capture technique (a DNA/RNA-hybridisation assay) is quick and easy to perform, but has a low clinical sensitivity. Therefore, it is generally recommended to use DNA/RNA amplification techniques, such as PCR. There are excellent PCR-primer sets available which identify all relevant oncogenic HPV types. These tests are also commercially available, often in combination with associated detection equipment. The PCR reactions are often in a multiplex form of High Risk HPV types (often combined with ELISA-based detection techniques). Because characterization of the HPV type is of increasingly interest (e.g. for therapeutic vaccination), these multiplex PCR reactions are often combined with HPV typing methods like reverse dot/line blot methods (Fig. 3.21).

EBV

EBV is able to immortalize cells and is involved in the pathogenesis of certain malignant lymphomas, particularly Hodgkin's lymphoma and lymphomas in immune-deficient and transplantation patients. Morphologically, Hodgkin's lymphoma is highly similar to large cell anaplastic lymphoma. However, the latter is EBV-negative. Detection of EBV by means of in situ hybridization of the EBER1/2 RNA is therefore important for the differentiation between these two types of lymphoma (Fig. 5.3). EBV has also been implicated in the pathogenesis of a number of specific carcinoma types in the stomach and nasopharynx that are characterized by highly atypical epithelial cells (which are sometimes sparsely present) and many lymphocytic infiltrates, which are called 'lymphoepithelioma like' carcinomas. Detection of EBV is therefore important for the differentiation between these types of carcinomas and similar carcinomas as well.

HHV8

HHV8 is involved in the development of Kaposi's sarcoma, a vascular malignant tumour that occurs mainly in HIV-positive patients. Detection of HHV8 by PCR is helpful in differentiating between Kaposi's sarcoma and similar vascular tumours.

Specific infections

Detection of specific infections that are caused by difficult or non-culturable microorganisms is usually PCR-based. *Chlamydia trachomatis* infections of the female genital tract can cause infertility, and detection can therefore be done on biopsies of the cervix and endometrium.

Determination of the *M. tuberculosis* complex (*M. tuberculosis, M. bovis, M. microti, M. africanum*) is important for the choice of antibiotic therapy, which is

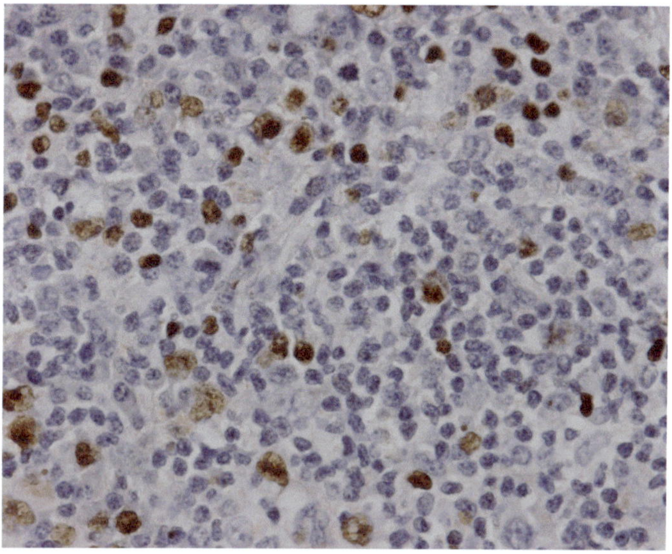

Fig. 5.3 Example of EBER 1/2 RISH on lymphoid tissue. The strong EBV-staining of blast cells is caused by an infectious mononucleosis (Pfeiffer's disease)

different for mycobacteria that do not belonging to this family. Other microorganisms, such as HSV (e.g. in liquor, sputum), CMV (retina, oesophageal), *Tropheryma whipplei* (Whipple's disease) and *Bartonella henselae* (cat scratch disease) can also be detected with PCR-techniques.

5.4 Molecular Diagnostics of Genetic Changes

Molecular diagnostics in pathology is focused on the diagnosis of tumours. Prior to illustrating the various applications, the various types of genetic changes in tumour cells will be described.

5.4.1 Types of Genetic Changes

There is a wide variety of genetic changes that can affect the functions of genes. These can either be chromosomal—or gene—alterations that may result in reduced, absent, or increased gene expression or the expression of an aberrant protein (Table 5.1). Chromosomal changes may give rise to a change in gene expression by displacement of the genes with respect to their promoters. Translocations involved

Table 5.1 The main molecular genetic changes at chromosome level and gene level and their effect on gene expression and protein production

Change	Effect
Chromosomal level	
Aneusomy/monosomy	Reduced gene expression
Polysomy	Increased gene expression
Translocation	Increased gene expression/expression oncogenic protein
Inversion	Reduced/increased gene expression or expression of an aberrant protein
Gene level	
Mutation	Reduced gene expression/expression aberrant protein
Insertion	Reduced gene expression/expression aberrant protein
Deletion	Reduced gene expression/expression aberrant protein
Amplification	Increased gene expression
Methylation (promotor)	Reduced gene expression

in tumour formation often result in aberrant gene expression by promotor rearrangement or give rise to the formation of fusion proteins.

Polysomy, monosomy and aneusomy can be detected using karyotyping, centromere fluorescence in situ hybridization (FISH), chromogenic in situ hybridization (CISH), silver in situ hybridization (SISH), MLPA or comparative genomic hybridization (CGH). Inversions and ring chromosomes can only be demonstrated with karyotyping. Translocations can be visualized by karyotyping and locus-specific FISH/CISH/SISH (Vol. 1, Chap. 8) or by RT-PCR (Vol. 1, Chap. 5) and, since recently, by NGS. Insertions, deletions and mutations can be detected with Sanger sequencing (Vol. 1, Chap. 8), NGS or qPCR. Amplifications can be analysed with MLPA, NGS and FISH or CISH. Methylation can be demonstrated by methylation specific PCR or related techniques, such as MLPA.

Pathology diagnostics makes use of short tandem repeats (STR) in the DNA for tissue identification, as described above, but also to identify STR instability (microsatellite instability: MSI). In addition to a prognostic marker, MSI serves as an indication for (germline) genetic defects in mismatch repair genes, which can give rise to various malignancies.

These STRs and the changes therein are identified with PCR, followed by fragment length analysis. Also, changes in gene expression belong to the field of molecular pathology, in particular molecular tumour pathology.

5.4.2 Molecular Genetic Changes in Tumour Cells

Some tumours arise from only a small number of specific changes. This is the case for some leukaemias, lymphomas and soft tissue sarcomas. Most tumours, however, arise as a result of a multi-step acquisition of changes in genes associated with proliferation, apoptosis, cell aging, angiogenesis, adhesion, invasion and metastasis, and immunological profile. Eventually, cells present as uncontrollably growing, selfish, invading and metastatic cells.

Genes involved in the onset of cancer (although somewhat simplified) are divided into two groups: the oncogenes and the tumour-suppressor genes:

1. Oncogenes are genes that, when activated, stimulate carcinogenesis. In principle, the activation of one of the alleles is sufficient for an oncogenic effect. Oncogenes are usually activated (for example amplified or mutated) normal genes (proto-oncogenes), but can also originate from translocations. It often concerns genes coding for growth factors, growth factor receptors, signal transduction proteins, transcription factors, cell cycle regulatory proteins and apoptosis inhibitors. Notable examples include MYC, RAS, EGFR, MDM2, CDK4 and HER-2/neu
2. Tumour-suppressor genes on the other hand, inhibit tumour growth and if their function decreases or is lost (for example by point mutations, deletions, or the methylation of promoters), this contributes to carcinogenesis. In principle, both alleles must therefore be affected. The classic model for this is the inactivation of one allele by a (sometimes germline) point mutation, followed by loss of (function of) the other allele (Knudsen-model). It often concerns genes encoding cell cycle regulatory proteins, transcription factors and apoptosis inducing proteins. Famous examples are p53 and the retinoblastoma protein.

Almost all malignancies are the result of multiple genetic changes in oncogenes and tumour suppressor genes. Sometimes these genetic changes occur in a specific sequence, but in general, carcinogenesis will be the result of a fairly random accumulation of a minimum number of major and minor genetic changes (the bingo principle). This gradual progression corresponds well with the morphological progression of normal (epithelial) cells via intermediate stages of hyperplasia, dysplasia and in situ carcinoma to invasive malignancy. Apart from specific translocations (as seen in lymphomas and soft tissue tumours), most malignancies are genetically heterogeneous. However, in some malignancies, genetic alterations can be recognized that are more or less specific. On this basis, molecular analysis can be helpful for the diagnosis and to specify the susceptibility to therapy. Combinations of genetic and epigenetic alterations eventually lead to changes in gene expression, which can be determined with expression microarrays. For many tumours, genetic (expression) signatures have been developed that predict prognosis and the response to therapy. An example is the Mammaprint, which can determine the probability of metastasis of breast cancer, thereby assisting the doctor in the choice of treatment (surgery with or without additional chemotherapy). The test measures the mRNA expression of 70 genes in the tumor tissue by microarray.

This expression profile is compared to a set of reference values (the reference profile), which is based on a group of 78 tumors with a known 10 year metastasis profile. Furthermore, these signatures can function as a tumour fingerprint to differentiate between different tumours (for example due to the metastasis of a known primary tumour, see Fig. 5.4).

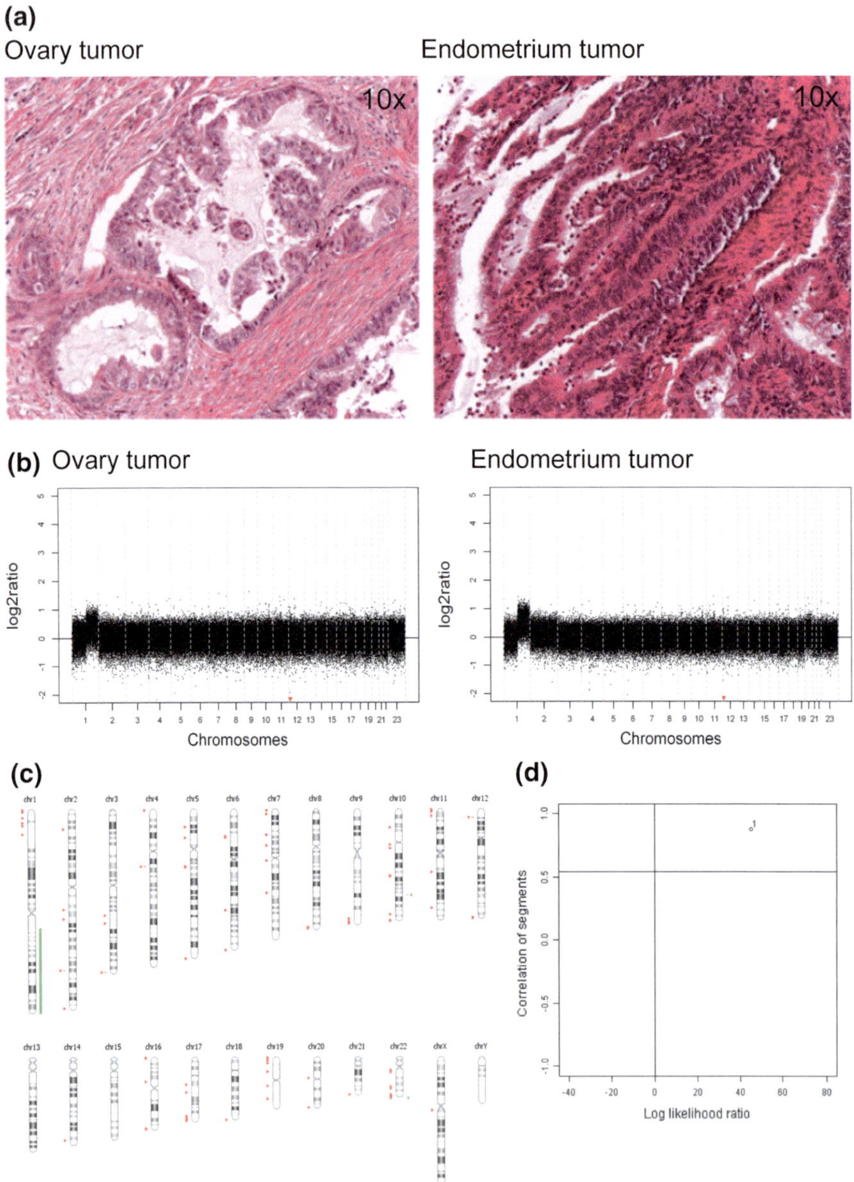

◀**Fig. 5.4** Comparative genomic hybridization (CGH) for determining clonal relationships between tumours. With CGH, so-called copy number changes (loss or gain) of the DNA can be detected between, for example, normal tissue and tumour tissue. For this purpose, an identical amount of normal DNA (stained with a red fluorescent label) and tumour DNA (stained with a green fluorescent label) are hybridized on a so-called CGH-microarray. This is a (glass) slide with hundreds of thousands of DNA-specified probes. In a competitive reaction some areas of the CGH-array will bind more (*green gain*) or less (*red gain*) tumour DNA compared to the normal DNA. If there is no difference in binding between the tumour and the normal DNA, a yellow fluorescent signal will be measured. With help of fluorescence microscopy and quantitative image analysis, the fluorescent ratios of each area are measured. Because each area on the CGH-array corresponds to a known DNA area, it can be determined which chromosomes exhibit a loss of or a gain. Most DNA losses and gains are unique events and form the "fingerprint" of a tumour. This fingerprint can be used to establish a possible common origin of different tumours. This example relates to a 65-year-old woman with a tumour in both the ovary and the endometrium. On the basis of histology (**a**) and P53-mutation analysis (no mutations present) both tumours were evaluated as two separate primary tumours. In the CGH-plots of both tumours (**b**) the fluorescent ratios (log2 ratio tumour/normal DNA) are plotted against the chromosome positions. Both of these tumours show an identical increased ratio for a portion of chromosome 1, caused by an amplification of the long-arm of chromosome 1 (**c**, *green bar*). The clonal relationship between both tumours is confirmed by a graphical plot, which combines two statistical tests to estimate the probability that both tumours are identical (the higher the score on both axes, the greater the agreement between both tumours) (**d**)

5.4.3 Pathological Indications

The following indications can be distinguished in molecular diagnostics of tumours:

1. Diagnostics to distinguish benign from malignant tumours.

In most cases, the pathologist can distinguish between benign and malignant tumoursusing light microscopy. Where needed, the pathologist can be aided by immunohistochemical, quantitative or electron microscopical techniques. However, this is difficult in a number of areas. This is the case for some lymphoid disorders, for which the detection of clonality (where all cells in a tumour are derived from a single progenitor cell) is an important diagnostic determinant. Obviously, this is crucial for the clinician to prepare an appropriate treatment plan.

2. Diagnostics to classify tumours.

Some tumours are difficult to classify, such as lymphoid tumours and tumours arising from bone and soft tissue. A good classification is, however, very important, since complementary therapies can vary widely, even when the tumours are morphologically very similar. Immunohistochemistry is sufficient in many cases, but the genetic profile (e.g. tumour specific translocations) is often crucial in specific cases, like soft tissue tumours.

3. Diagnostic tests to determine clonal relationships between two or multiple synchronous or metachronous biopsies of a patient.

In case a patient develops a second (metachronous) tumour, the question may arise whether this is a metastasis of the first or a second primary tumour. An example is an examination of the relationship between multiple squamous cell carcinomas of the (upper) respiratory tract, a relatively common situation. To differentiate between these tumours, p53 mutation, loss of heterozygosity (LOH) or CGH patterns, which function as genetic fingerprints, are compared. The same naturally applies with two simultaneous (synchronous) tumours (Fig. 5.4).

4. Detection of micrometastases and 'minimal (residual) disease'.

Accurate sampling is a key principle within histopathology in order to achieve adequate effectiveness. From an average lymph node for example, approximately 2500 standard sections are cut. Therefore, the entire assessment of a lymph node is impossible, in practice. Normally, a limited number of slides are taken throughout the lymph node at regular intervals. Although daily practice teaches that this has a high yield, smaller metastases can easily be missed. In addition, it is sometimes very difficult to recognize a single tumour cell in a background of many benign cells (such as in the lymph nodes and at resection margins), which can have important consequences. With sensitive PCR techniques, such micro-metastases can be detected, making use of tumour-specific changes, such as the expression of certain genes, mutations or translocations.

5. Prognosis, determination and prediction of response to therapy.

Even if it is determined that a tumour is malignant, its clinical behaviour may still vary considerably. Some tumours act indolent, while others are very aggressive and possibly need additional therapy. Genetic changes in a variety of tumours have been described that are either associated with a specific prognosis or that can predict the response to a specific therapy. Especially the latter is becoming more and more important with the increasing availability of so called "targeted therapies". Mutation analysis (and possibly gene expression profiles, see Sect. 5.4.2) is most important in this.

The indications listed will be explained in more detail in the following sections by means of examples. These will show that significant variations in approach exist for different tumours. The different strategies are linked to the nature and complexity of the lesion, with the current level of knowledge with regard to the genetic changes in specific lesions, the presence or absence of other independent (for example, immunohistochemical) characteristics and in general, to the presence of a 'gold standard' to which molecular findings should be related. Developments in Molecular Pathology go very fast. New (RT-)PCR or FISH tests are developed regularly to detect variant breakpoints in sarcomas, some with potentially important diagnostic meaning. The same applies to markers for the purpose of detecting micrometastases and minimal (residual) disease. However, the biggest change in molecular pathology is due to the emergence of NGS, which enables the simultaneous detection of multiple mutations and translocations in dozens of genes.

5.4.4 T/B Cell Clonality

Lymphomas are malignant tumours that derive from lymphoid T- and B-cells, usually in the lymph nodes. Molecular diagnostics of malignant lymphomas is mainly a specialty of the pathologist, while acute and chronic leukaemia and other bone marrow abnormalities (myelodysplastic and myeloproliferative disorders) are mainly the area of the haematologist (see Chap. 6). In almost all malignant B- and T-cell lymphomas, clonality can be detected by PCR. These clonalities are usually determined by using commercially available multiplex-PCR reactions. The probes that are used in these kits have been developed in the European Biomed 2 program. In addition, B- and T-cell malignancies can be characterized in more detail on the basis of the number of tumour-specific translocations. These translocations can be determined with FISH, using "break-apart" or "fusion" probes (Sect. 5.4.5), or with the aid of (RT-)PCR or NGS.

5.4.5 Molecular Diagnosis of Chromosomal Deviations

Changes in chromosome number may be genetic (cytogenetic conditions) or autosomal (tumours). Traditionally, cytogenetic changes are examined by karyotyping of dividing cells. These chromosomal abnormalities can now also be determined in tissues using MLPA, FISH/CISH and array-CGH.

Cytogenetic changes with a genetic origin

In pathology, detection of chromosomal abnormalities is used for the determination of *mola hydatidosa* and abnormalities in new-borns or miscarriages. In the last situation, this mainly concerns trisomies of chromosomes 21 (Down's syndrome), 13 (Patau syndrome) and 18 (Edwards syndrome). CISH/FISH and MLPA are the most appropriate methods to detect these abnormalities. Other chromosomal abnormalities occur, but are less frequent and usually lethal.

Complete and partial *mola hydatidosa* represent 80% of the trophoblast diseases during pregnancy. These are (pre-malignant) disorders of the placenta, which are often characterized by abnormalities in the number of sex chromosomes. In case of a complete mola, the empty oocytes are fertilized by one or two sperm cells, and thus only contain paternal genetic material (46XX or 46XY). In partial mola, an egg is fertilized by two sperm cells and these are therefore mostly triploid: 69XXX, 69XXY or 69XYY. The detection is usually done by FISH or CISH for X and Y-chromosomes, but can also be done with conventional DNA cytometry.

Cytogenetic changes in tumours

Chromosomal abnormalities are common in tumours. However, in most cases these abnormalities are not of diagnostic value. A good example of a diagnostic application based on chromosomal aberrations is the UroVysion test for bladder carcinomas. These cancers are usually diagnosed with urine cytology, which is not very specific. The UroVysion-test is a 4-color FISH, which tests for loss of 9p21 (an area with tumour suppressor genes) and aneusomy of chromosomes 3, 7 and 17 (Fig. 5.5). FISH is performed on cytological material. The pattern of the four signals is compared to the normal situation and deviations are evaluated and classified according to a scoring table. This test shows a higher diagnostic specificity for the detection of bladder carcinomas than conventional methods. Another interesting diagnostic application based on chromosomal aberrations is array-CGH, which can distinguish between sporadic and BRCA1 and BRCA2 related breast cancers (see Vol. 1, Chap. 8).

Translocations

Chromosomal translocations occur in 20% of all tumours and play an important role in the initiation of oncogenesis. The fusion genes that are generated can result in the overexpression of one of the involved genes or in the formation of hybrid genes. This can lead to the activation of transcription factors, which is the case in many lymphomas and sarcomas, among others. Translocations are often tumour-specific, making them ideal diagnostic markers. Several techniques are suitable to detect

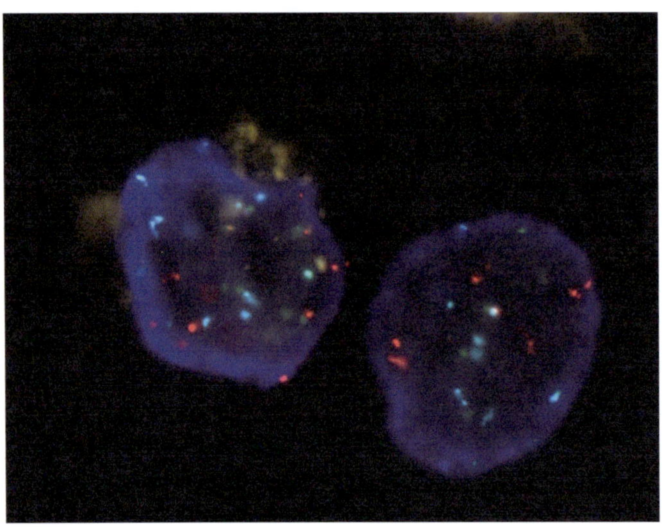

Fig. 5.5 UroVysion FISH. Example of FISH UroVysion, wherein polysomy (more than two copies per nucleus) of chromosomes 3 (*red*), 7 (*green*) and 17 (aqua) is visible in two malignant urothelial cell nuclei (*blue*). The right-hand nucleus also shows a loss of loci 9p21 on chromosome 9 (*gold*)

these translocations: reverse transcriptase-PCR (RT-PCR), FISH and classic cytogenetics (see Vol. 1, Chaps. 4 and 5).

Translocation analysis with FISH is performed using "break-apart" probes or "fusion" probes. The latter uses different fluorescent probes that bind to each of the fusion partners. When a fusion gene is present, both probes "melt" together, resulting in a colour overlap between both probes. The advantage of this technique is that a specific translocation can be demonstrated. The disadvantage is that false-positive translocations can be observed in overlapping cells. With "break-apart" probes, one probe is located upstream of the breaking point and another probe is positioned downstream of the breaking point of one of the involved genes. In case of a translocation, both probes will become visually separated (Fig. 5.6). Although the risk of false-positivity is smaller, the specific fusion partner cannot be identified.

FISH is a rapid technique that is equally suitable for fresh (frozen) and formalin-fixed paraffin-embedded material. The limited availability of (expensive) commercial probes is a limiting factor, but more and more are becoming available.

In RT-PCR, RNA is isolated and subsequently transcribed into c-DNA, which is thenamplified using two specific primers chosen on both fusion partners. A PCR-product will only be generated in the presence of a fusion transcript (See Vol. 1, Chap. 6 and Sect. 3.5.3). The advantage is that specific translocations can be demonstrated. In contrast to fresh (frozen) material, isolation of RNA from formalin-fixed paraffin-embedded material is difficult and sometimes impossible.

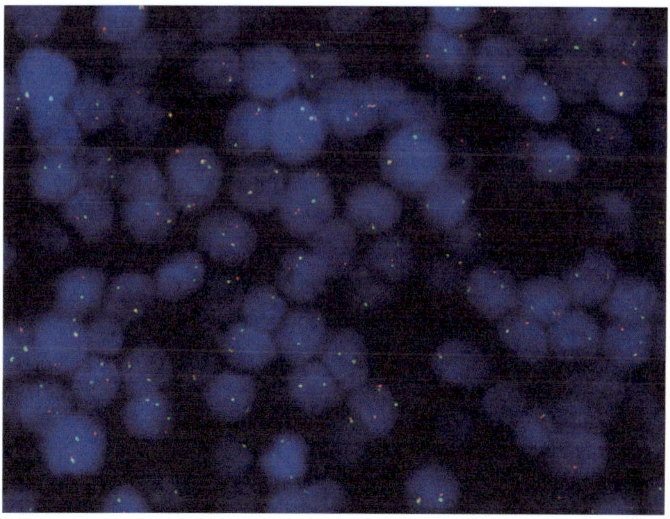

Fig. 5.6 "Break-apart" FISH Follicular lymphoma (100× magnification). This technique uses two fluorescently labelled probes (*red* and *green*), that bind to either side of the BCL2 gene (on chromosome 18). During a translocation, a fracture occurs in the BCL2 gene, whereby a part, including the green label, moves to chromosome 14. Apart from an intact wild type copy of BCL2 (*red* and *green* probes overlap, partly *yellow*), two separate probes are visible

The choice between the different techniques will depend on the question and the available material. When fresh (frozen) material is available, RT-PCR is preferred, while FISH is the best option for paraffin-embedded material. There has been little progress in the development of CISH and SISH for these applications.

Sarcomas

Sarcomas are malignant tumours originating from mesenchymal cells of the soft tissues (muscle, fat, connective tissue, nerves, vessels) or bones. The diagnosis of these tumours is difficult. Morphologically, there is much overlap between benign and malignant variants and the classification of malignant tumours is often difficult, even by means of immunohistochemistry (Fig. 5.7). Characteristic translocations leading to oncogenic fusion proteins that can help to diagnose the tumour have been described in a variety of bone and soft tissue sarcomas (Table 5.2). The group of round cell tumours (such as Ewing sarcoma, rhabdomyosarcoma, round cell liposarcoma, neuroblastoma) and the group of spindle cell tumours (including synovial sarcomas and malignant peripheral nerve sheath tumours) are of special interest for further molecular diagnostics. With the aid of translocation analysis, many different translocations can be detected, with direct implications for the choice of therapy.

Lymphomas

Lymphomas are malignant tumours originating from lymphoid B- and T-cells, usually in lymph nodes (see also Sect. 5.4.4). In certain lymphomas, a gene fusion

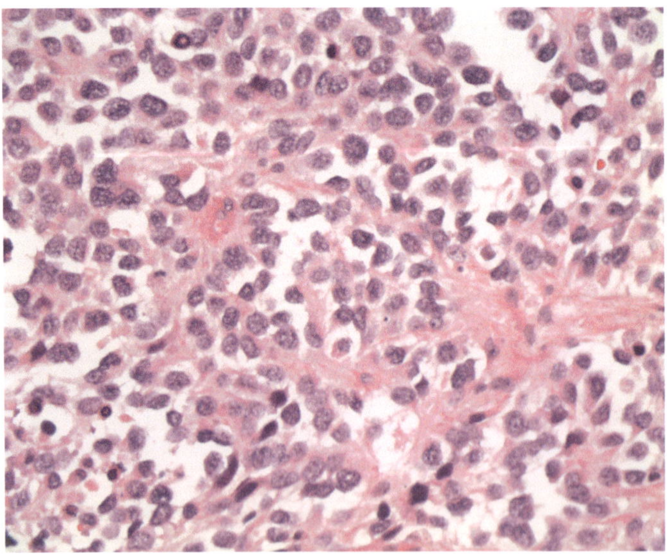

Fig. 5.7 Haematoxylin-eosin staining Ewing's sarcoma (10× magnification). Ewing's sarcoma, built up from small fields of monotonous round cells often with barely recognizable cytoplasm

Table 5.2 Characteristic translocations in various bone and soft tissue sarcomas and lymphomas

Clinical indication	Cytogenetics	Gene-fusion
Sarcomas		
Ewing/PNET	t(11;22)(q24;q12)	EWS-FLI1
	t(21;22)(q22;q12)	EWS-ERG
	t(7;22)(p22;q12)	EWS-ETV1
Alveolar rhabdomyosarcoma	t(2;13)(q35;q14)	PAX3-FKHR
	t(1;13)(p36;q14)	PAX7-FKHR
Myxoid liposarcoma	t(12;16)(q13;p11)	FUS-CHOP
	t(12;22)(q13;q12)	EWS-CHOP
Clear cell sarcoma	t(12;22)(q13;q12)	EWS-ATF1
Myxoid chondrosarcoma	t(9;22)(q31;q12)	EWS-CHN
Desmoplastic small round cell tumour	t(11;22)(p13;q12)	EWS-WT1
Synovial sarcoma	t(X;18)(p11;q11)	SYT-SSX1/2
Lymphomas		
Burkitt's lymphoma	t(8;14)(q24;q32)	IGH-MYC
Follicular lymphoma	t(14;18)(q32;q21)	IGH-BCL2
Anaplastic large cell lymphoma	t(2;5)(p23;q35)	NPM1-ALK
Mantle cell lymphoma	t(11;14)(q13;q32)	IGH-BCL1
MALT lymphoma	t(11;18)(q21;q21)	API2-MALT

between an immunoglobulin (IGH) gene and a transcription factor gene occurs, wherein the so-called IGH enhancer elements result in the (over)expression of the transcription factor. Many malignant B- and T-cell lymphomas can be detected on the basis of more or less specific tumour translocations (Table 5.2, Fig. 5.6).

Lung carcinomas

Lung tumours in particular are characterized by specific translocations that activate oncogenic signalling pathways. The fusion of the ALK (Anaplastic lymphoma kinase) gene with the EML4 gene, results in the over expression of the ALK protein, resulting in the activation of major signalling pathways (MAPK and PI3 K/AKT) that are directly involved in oncogenesis. This is of great clinical importance, since it can be specifically inhibited by ALK specific tyrosine kinase inhibitors (TKIs). The same holds true for the ROS protein that is also upregulated after translocation and is also sensitive to inhibition with TKIs.

Loss of 1p19q

The loss of both the short arm of chromosome 1 and the long arm of chromosome 19 is characteristic for oligodendrogliomas. After glioblastomas, these are the most common primary brain tumours. Oligodendrogliomas differ from other gliomas by their relatively high chemosensitivity (e.g. temozolomide) and thus have a relatively good prognosis. The 1p19q status is therefore an important diagnostic marker. There are several methods for the detection of 1pq19 loss: LOH analysis, FISH and

MLPA. In the case of 1p and 19q loss on one of the chromosomes, specific signals of the 1p and 19q-specific probes will be halved. The probes are directed against multiple locations of 1p and 19q, which ensures a good detection of the complete loss of 1p19q. Additional diagnostics provides more information about the cytostatic sensitivity of gliomas. MGMT promoter methylation is an important predictor for temozolomide sensitivity in glioblastomas (see Sect. 5.4.9).

5.4.6 Molecular Diagnostics of Gene Amplification

Many tumours are characterized by the amplification of one or more proto-oncogenes. Thanks to a disturbed replication of the DNA, multiple copies of the gene are present, which results in the overexpression of mRNA and protein. Amplifications can be informative with respect to the prognosis, tumour identification and the choice of treatment. Various techniques can be applied for the detection of gene amplifications, including FISH/CISH/SISH/MLPA and NGS.

ERBB2 (HER-*2/neu) amplification*

HER-2/neu is a proto-oncogene which encodes a transmembrane tyrosine kinase growth factor receptor belonging to the epidermal growth factor receptor family. The coding gene, *ERBB2,* is amplified in 10–20% of breast cancers and is associated with a poor prognosis. However, the recently developed drug trastuzumab, a humanized monoclonal antibody against the extracellular domain of HER2, offers good opportunities for the treatment of women with HER2-positive breast cancer. Therefore, determination of the HER2 status is important to select patients for treatment with trastuzumab. An immunohistochemical staining with HER2 antibodies can be used for a semi-quantitative estimation of the HER2 expression. FISH, CISH (Fig. 5.8), SISH or MLPA may be used (Fig. 5.9) in order to demonstrate *ERBB2* gene amplification.

CDK4 *and* MDM2 *amplification*

When identifying sarcomas (see Sect. 5.4.5), liposarcomas are difficult to distinguish from lipomas. In contrast to lipomas, liposarcomas are characterized by the amplification of *MDM2* and *CDK4*. Although *MDM2*/*CDK4* amplification is not totally exclusive for liposarcomas, this is still a strong criterion for the diagnosis of liposarcomas. MLPA is ideal for the simultaneous detection of both gene amplifications.

5.4.7 Microsatellite Instability

The DNA mismatch repair system is responsible for the recovery of all base-base errors (mismatches) and small insertion/deletion errors that may arise during

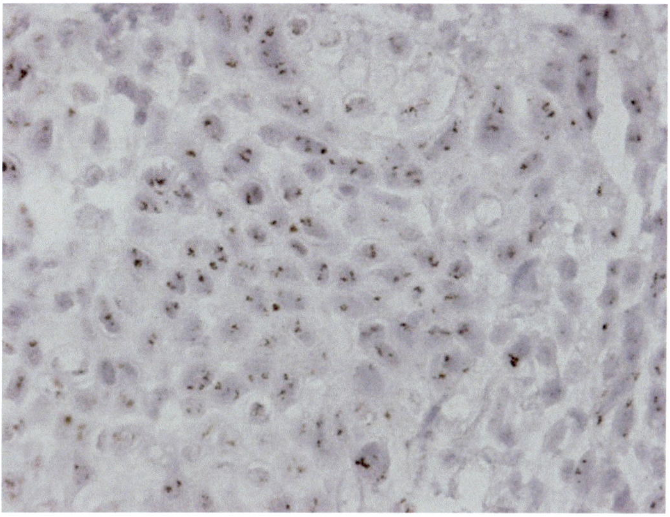

Fig. 5.8 ERBB2 CISH-breast cancer (10× magnification). In most of the cells, more than five ERBB2 copies are visible, which indicates a ERBB2 amplification

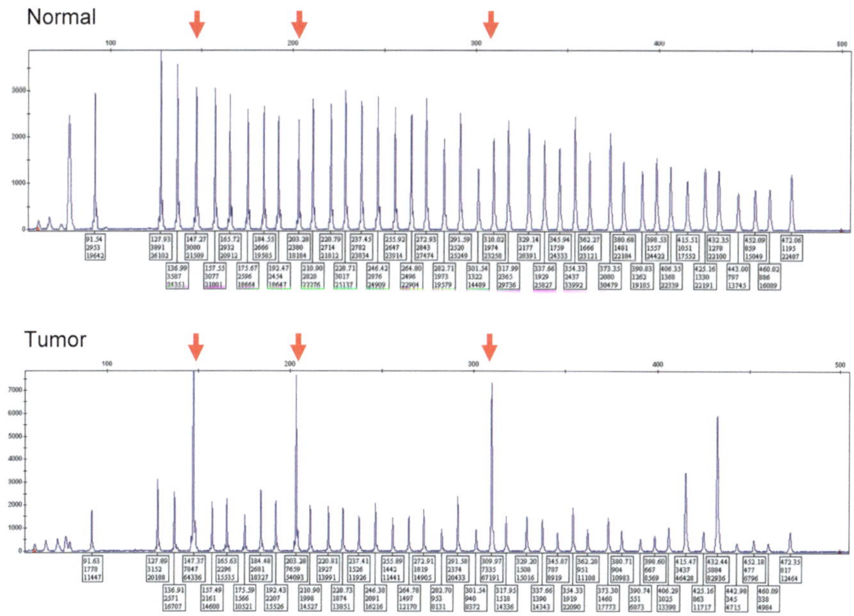

Fig. 5.9 ERBB2 amplification in breast carcinoma detected by MLPA. Of the forty probe sets in this particular MLPA, there are three specifically targeted against the HER2 gene (*red arrows*). All probe sets are then amplified with generic fluorescently-labelled PCR-primers and analysed by means of fragment analysis. In the obtained genescan, the fluorescence is plotted against the various fragment sizes associated with the various probes. The measured amount of fluorescence is then related to the number of gene copies. The upper row GeneScan shows an example of a tumour without HER2 amplification. The lower row GeneScan shows an example of a tumour with HER2 amplification, wherein the fluorescence for the HER2-probes is increased nearly four times

replication of the DNA in a cell. The different mismatch repair proteins form complexes that successively recognize, delete and restore DNA errors. Failure of this system results in genetic instability at the nucleotide level. This is called microsatellite instability (MSI) because errors first occur in repetitive DNA sequences, called microsatellites. MSI is defined by a change in length of a micro-satellite, either by insertion or deletion of repeating units. MSI is determined by PCR amplification of microsatellites at different loci, which are compared between normal and tumour tissue (Fig. 5.10). When more than 30% of the loci show instability, the tumour is marked as MSI.

For gastro-intestinal tumours, it is important to distinguish between sporadic and hereditary tumours, such as Lynch syndrome. Hereditary tumours affect the patient, but also the family of the patient. Lynch syndrome patients are carriers of a genetic mutation in one of the DNA mismatch repair genes (*MLH1*, *MSH2*, *MSH6* or *PMS2*), which causes MSI. While sporadic tumours only show MSI in 15% of the cases, MSI is an important indication for Lynch syndrome. The diagnosis is often supported by the absence or presence of immunohistochemical staining of the

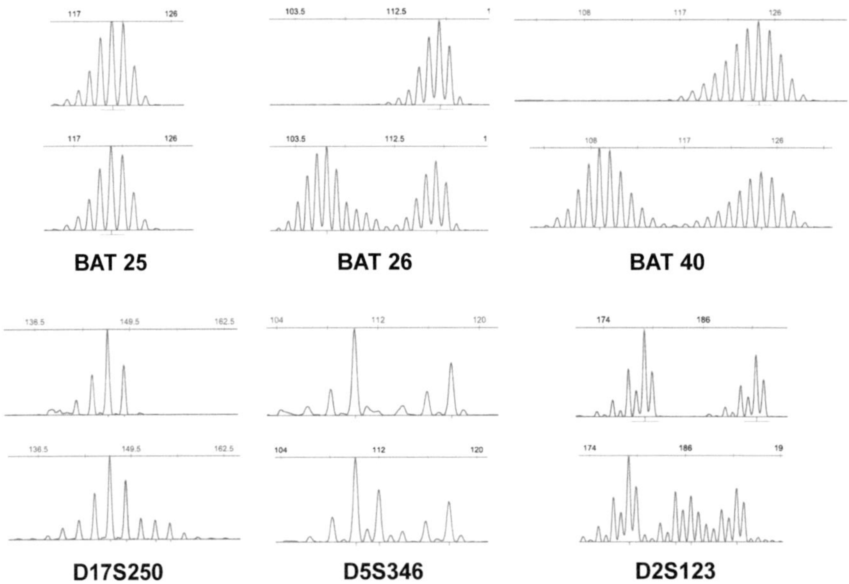

Fig. 5.10 MSI analysis. MSI analysis of 6 different loci, three mononucleotide markers (BAT 25, 26 and 40) and three dinucleotide markers (D17, D5, and D2), by means of repeat PCR (GeneScan). The loci of the tumour tissue (*below*) are compared with the normal tissue (*top*). The BAT 25 locus (homozygous) and the D2 locus (heterozygote) show no difference in fragment length between the tumour and the normal tissue and are microsatellite stable. In case of the BAT 26 and 40 loci, one of the alleles lost a number of nucleotides. Whereas in the D17 and D5 loci, one of the alleles has just acquired a number of nucleotides. In both cases we refer to microsatellite instability. In total, 4 out of 6 loci are microsatellite unstable, and one speaks of a microsatellite unstable tumour

mismatch repair proteins. Furthermore, MSI in sporadic gastro-intestinal tumours is most often caused by promoter hypermethylation of *MLH1*, often accompanied with a mutation in the *BRAF* gene. Both are therefore important additional diagnostic markers (see Sects. 5.4.8 and 5.4.9).

5.4.8 Mutation Analysis

Mutation analysis in pathology serves several purposes. For example, it is used for the detection of clonal relationships between two or more synchronous or metachronous biopsies of a patient (see Sects. 5.4.3 and Nyati et al. 5). Some mutations have a prognostic value and more and more mutations are correlated to a so-called "targeted" therapy. These involve small molecule inhibitors (monoclonal antibodies or TKIs), which specifically inhibit a mutated protein (amplified or (in)activated). The target proteins are mostly crucial proteins in a signal transduction pathway. The epidermal growth factor receptors (EGFR), including EGFR and HER-2/neu and their underlying signal transduction pathways, such as the MAPK and PI3 K/AKT pathways, are the main targets for this new generation of cytostatics (Fig. 5.11).

Several techniques are available for mutation analysis. If there is a specific mutation site (hotspot), then mutation-specific (quantitative) PCR is a fast and sensitive method. High resolution melting (HRM) is a relatively new technique, which is very well suited for the determination of mutations located in hotspot areas. HRM is a post-PCR analysis method based on detecting small differences in PCR melting (dissociation) curves (Fig. 5.12). In the case of random mutations, sequence analysis is the preferred technique (Fig. 5.12). There has been tremendous progress in the development of next-generation sequencing techniques for mass sequencing of large numbers of genes (Fig. 5.13). Because NGS can analyse various DNA mutations in several genes simultaneously, it is ideal for mapping the different tumour routes and for assessing possible (targeted) treatments. In principle, all tests can be performed on both DNA and RNA (cDNA).

p53

p53 is the most mutated tumour-suppressor gene in tumours and plays a central role in the signalling cascade that regulates cell growth and apoptosis. In the case of cell damage, p53 stops cell growth (in order for the DNA damage to be repaired), or induces apoptosis. Because almost all tumours develop unique p53 mutations, mutation analysis of p53 is very suitable for the detection of clonal relationships between two or more tumours of a patient. One example is the determination of the relationship between multiple squamous cell carcinomas of the (upper) respiratory tract, a relatively common situation. Mutation analysis of p53 is focused on the exons that are most frequently mutated (exons 5 through 8). P53-mutation analysis has an increasing diagnostic and therapeutic significance.

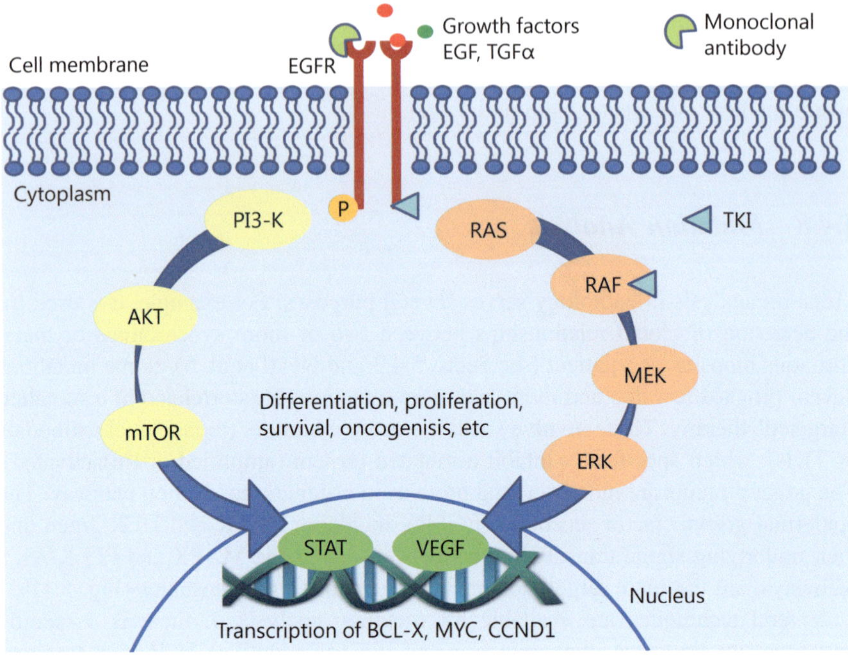

Fig. 5.11 EGFR signalling pathway in tumours. Epidermal growth factor receptor (EGFR) exists on the cell surface and is activated by binding of its specific ligands, including epidermal growth factor and transforming growth factor α (TGFα). Upon activation by its growth factor ligands, EGFR undergoes a transition from an inactive monomeric form into an active homodimer. EGFR dimerization stimulates its intrinsic intracellular protein-tyrosine kinase activity. As a result, autophosphorylation of several tyrosine (Y) residues in the C-terminal domain of EGFR occurs. This autophosphorylation elicits downstream activation and signalling by several other proteins that associate with the phosphorylated tyrosines through their own phosphotyrosine-binding domains. These downstream signalling proteins initiate several signal transduction cascades leading to DNA synthesis and cell proliferation. Activating mutations in EGFR, RAS, RAF and PI3-K cause autophosphorylation independent of upstream activation and result in uncontrolled cell proliferation. This can be countered by the use of so-called tyrosine kinase inhibitors (TKI's) which block the tyrosine kinase domain and prevent phosphorylation by mimicking ATP. In case of EGFR (or Her2-neu), amplification monoclonal antibodies can be applied to prevent the activation of the signalling pathway

EGFR

The epidermal growth factor receptor (EGFR) is the starting point of a number of major signalling pathways (MAPK and PI3K/AKT) that are involved in oncogenesis and it is overexpressed in many carcinomas (Fig. 5.11). In colon tumours, (over)expression of EGFR can be inhibited by monoclonal antibodies (such as cetuximab).

Lung carcinomas, which normally show changes in the tyrosine kinase domain of EGFR, are usually treated with TKIs. The response to these TKIs strongly

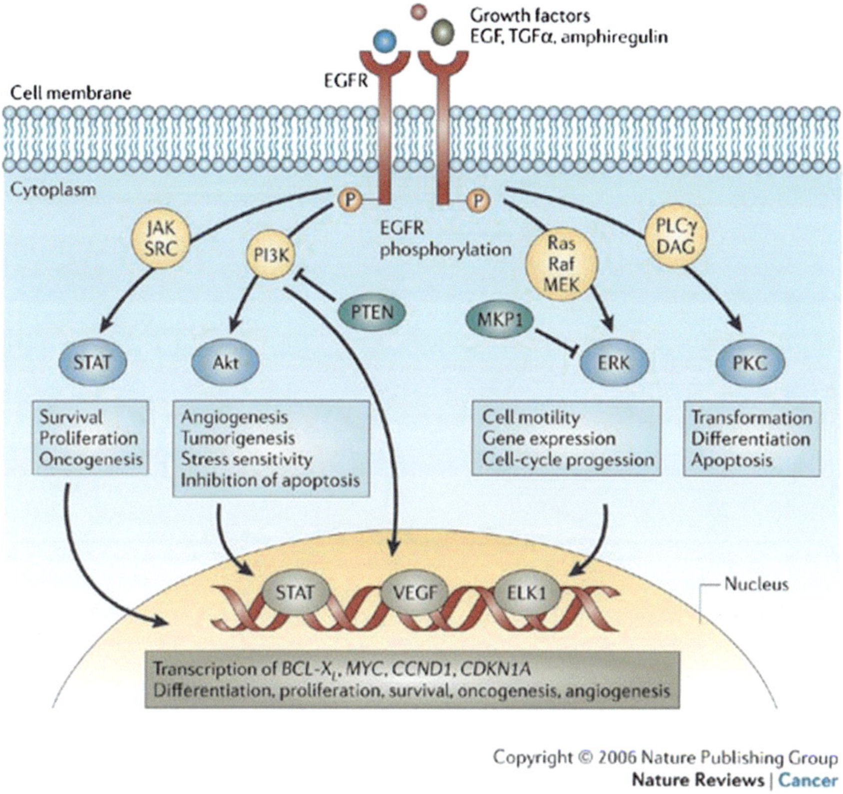

Fig. 5.11 (continued)

increases in the presence of activating EGFR mutations, which can be found in mutation hotspots on exons 18, 19 and 21, THI sensitivity decreases when mutations in exon 20 are present (Fig. 5.13).

RAS

KRAS and NRAS mutations are found in various tumours, promoting uncontrolled cell growth through various effector pathways. Most RAS mutations are activating point mutations located in exons 2 through 4. Because KRAS and NRAS are located downstream of EGFR in the MAPK signal transduction pathway, RAS mutations are a negative predictor for anti-EGFR therapies, such as monoclonal antibodies and TKIs, for the treatment of colon and lung tumours, respectively.

BRAF

As an important link in the MAPK signal transduction pathway, BRAF mutations play a major role in a variety of tumours. Activating *BRAF* mutations are located in exon 15. Most commonly, the Valine at position 600 in *BRAF* is mutated to a

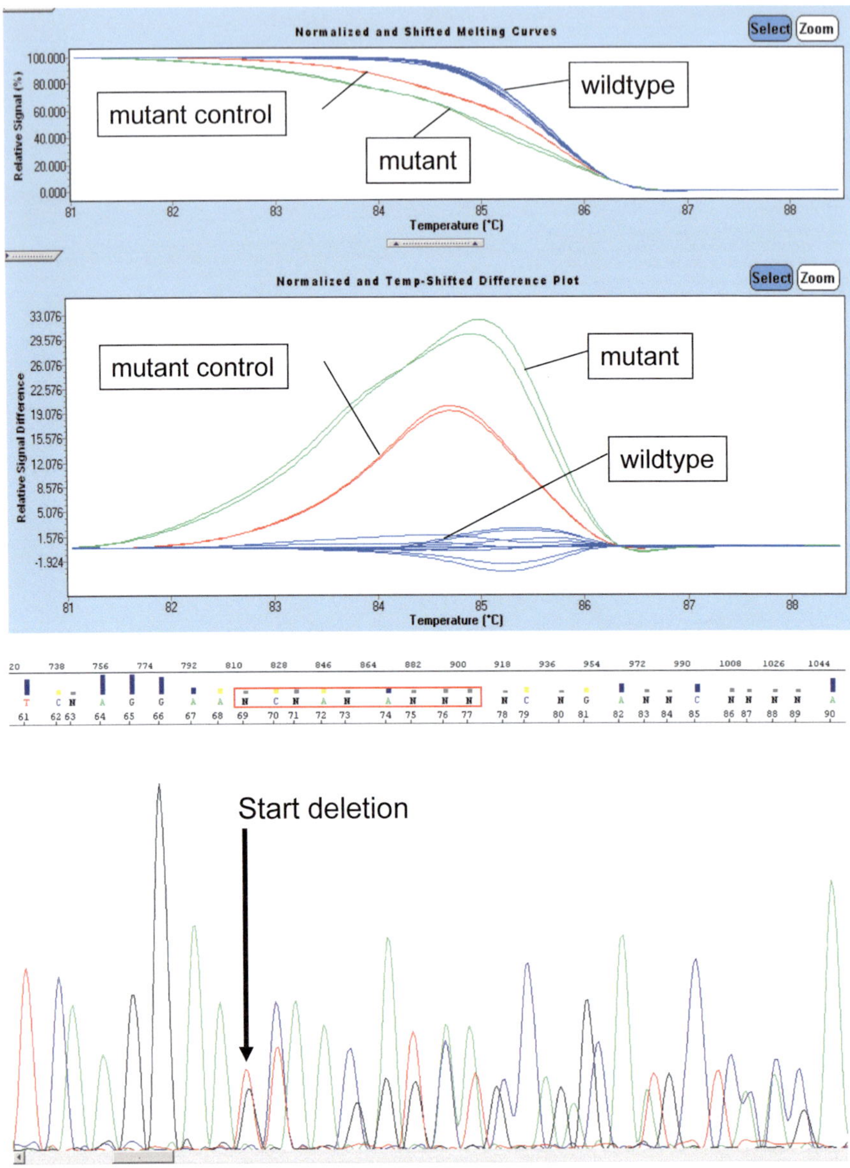

Fig. 5.12 EGFR mutation analysis using "High Resolution Melting" (HRM). The upper graph shows the melting curves of EGFR PCR products of different samples. The PCR products from samples with a heterozygous EGFR mutation (mutant) form less stable duplexes (mutant with wildtype) and have a lower melting temperature than PCR products of samples without EGFR mutation (wildtype), which form more stable duplexes. In the 'difference plot' (the second derivative of the melting curve) in the middle graph, the difference in melting temperature is much clearer. In order to confirm the mutation, the mutant PCR-product is analysed by Sanger sequence analysis (*bottom panel*). The sequence of the mutant allele is out of sync due to a deletion of nine base pairs (*red box*)

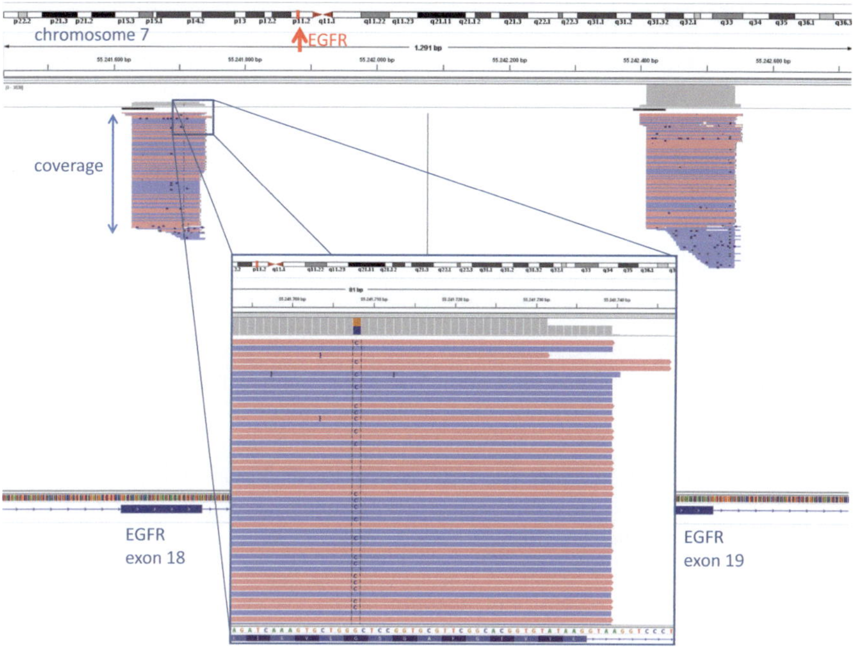

Fig. 5.13 EGFR mutation analysis using next generation sequencing. NGS is increasingly used in diagnostics for mutation analysis of genes. This can be achieved by applying a targeted approach, where only the clinically relevant genes or mutation hotspots are analysed. This can be done by performing a multiplex PCR with hundreds of PCR primers for dozens of genes. The resulting PCR amplicons are then sequenced on a NGS platform. This figure shows a screenshot of exons 18 and 19 of the EGFR gene using the Integrative Genome Viewer (IGV). Each individual sequence (read) is depicted by a pink bar (forward direction) or a purple bar (reverse direction), whereas the total of individual sequences describes the coverage (sequence depth) of a particular DNA section. In this example, a heterozygous G > C mutation in exon 18 was present in about half of the sequence reads

Glutamic Acid (V600E). Like *KRAS* and *NRAS*, *BRAF* is located downstream of *EGFR* and negatively influences EGFR directed therapy. However, upon treatment with specific BRAF inhibitors (such as sorafenib), these mutations have a positive predictive value for therapy. This applies in particular for melanomas, where over 60% of the tumours have an activating BRAF mutation.

In addition to the prediction of therapy, *BRAF* mutations also have an important role in the diagnosis of Lynch syndrome. Gastrointestinal tumours associated with Lynch syndrome are characterized by MSI (see Sect. 5.4.7 and Fig. 5.10). However, 15% of sporadic tumours also show MSI. In most cases, this is caused by promoter methylation of the *MLH1* mismatch repair gene. *MLH1* methylation is usually accompanied by *BRAF* mutations. Thus *BRAF* mutation, in addition to *MLH1* methylation, is a good indicator for sporadic gastro-intestinal tumours.

KIT and PDGFRA

Most gastrointestinal stromal tumours (GIST) contain an activating mutation in the tyrosine kinase receptor, KIT, or in the platelet-derived growth factor-receptor (PDGFRA). These tumours can be treated successfully with tyrosine kinase inhibitors. Success of treatment not only depends on the presence/absence of a mutation, but also on the position of the mutation. While tumours without mutations have a poor response to TKIs, tumours with KIT mutations in exon 11 show a strong response to TKIs.

Mutation detection in blood (liquid biopsies)

Blood contains very low concentrations of so-called cell-free DNA, which is shed in the blood after a cell has died (Fig. 5.14). This cell-free DNA is therefore a representation of all cells that have died, including tumour cells. Blood bound DNA fragments can be analysed for specific mutations using very sensitive techniques, such as droplet or digital PCR. The high sensitivity of digital PCR is achieved by dividing the sample into very small quantities (droplets) on which individual qPCR reactions are performed. A droplet carries either a mutation (digital: 1) or no mutation (digital: 0). This could serve as an alternative detection method when it is not possible to take a biopsy, but it could also serve as a tool to monitor the treatment of the tumour and detect potential resistance mutations at an early stage. If this technology is sufficiently matured it could even be used as a screening tool in the future.

5.4.9 Hyper Methylation Analysis

Many tumours originate from abnormal methylation patterns of tumour-suppressor genes and oncogenes. If the promoter of a tumour-suppression gene is hypermethylated, its DNA expression is blocked. Because these aberrant methylation patterns often occur at a very early stage of oncogenesis, its early diagnosis could have a role in early cancer detection or even in cancer prevention (see Vol. 1, Chap. 8).

MLH1 hypermethylation

MLH1 is a mismatch repair protein. When complexed with other mismatch repair proteins, it is able to repair DNA abnormalities. Deactivation results in MSI, an important marker for the classification of gastro-intestinal tumours (see Sect. 5.4.7, Fig. 5.10). In Lynch syndrome, this is caused by germline mutations in mismatch repair genes, including *MLH1*. In sporadic gastrointestinal tumours MSI is most often caused by promoter hypermethylation of *MLH1*. Therefore, the determination of the *MLH1* methylation status has an important role in the diagnosis of Lynch Syndrome.

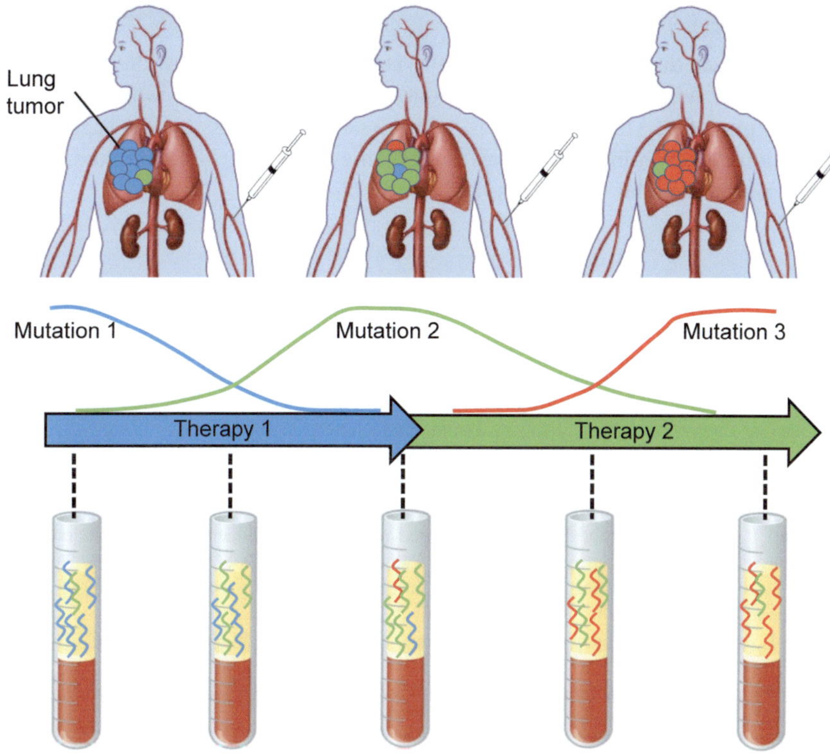

Fig. 5.14 Mutation detection in blood. In this example, a lung tumour is biopsied and genotyped before initiation of therapy. Blood-based analysis is performed on the cell-free DNA obtained from the blood serum. The patient is started on therapy 1, on the basis of the molecular analysis (Mutation 1) of the tumour. The response to therapy can be monitored regularly by analysing the cell-free DNA. The predominant tumour clone 1 (*blue*, Mutation 1) which is present at the start of therapy diminishes during therapy 1 (as indicated by a reduction in the numbers of this particular clone), while clone 2 (*green*, Mutation 2) is selected for during therapy (as indicated by a rise in the numbers of this particular clone). As a result, treatment is switched to therapy 2, which targets clone 2 (Mutation 2), and so on. The therapy can be adjusted in time according to the measured response in the blood

MGMT hypermethylation

MGMT encodes the DNA repair protein O6-alkylguanine-DNA-alkyltransferase (AGT), which protects the DNA against alkylating cytostatics. In principle, the absence of MGMT is a poor prognostic indicator in tumours, because it results in the accumulation of mutations and chromosome instabilities. However, the absence of *MGMT* makes a tumour very vulnerable for chemotherapy with alkylating agents (e.g. temozolomide). When treatment with alkylating cytostatics is considered, it is important to know the functional status of *MGMT*. Hyper-methylation of the promoter is a common cause for the elimination of *MGMT,* especially in gliomas, and serves as an important predictor for temozolomide sensitivity (see Sect. 5.4.5).

Methylation as a screening test

Due to the fact that promoter hypermethylation is a relatively widespread phenomenon and has an early onset, its detection is in principle well suited for screening purposes. Much research is being done into the possible applications of methylation detection in tumour screening, such as screening in nipple fluid, urine, sputum, cervix samples and faeces, but also in blood.

Recommended Literature

Agagnostopoulos I, Hummel M. Epstein-Barr virus in tumours. Histopathol. 1996;29:297–315.

Blaheta HJ, et al. Lymph node micrometastases of cutaneous melanoma: increased sensitivity of molecular diagnosis in comparison to immunohistochemistry. Int J Cancer. 1998;79:318–23.

Boland CR, et al. A National Cancer Institute Workshop on Microsatellite Instability for cancer detection and familial predisposition: development of international criteria for the determination of microsatellite instability in colorectal cancer. Cancer Res. 1998;58:5248–57.

Bubendorf L, et al. Multiprobe FISH for enhanced detection of bladder cancer in voided urine specimens and bladder washings. Am J Clin Pathol. 2001;116:79–86.

Cox JT, et al. History of the use of HPV testing in cervical screening and in the management of abnormal cervical screening results. J Clin Virol. 2009;45:S2–12.

Crowley E, et al. Liquid biopsy: monitoring cancer-genetics in the blood. Nat Rev Clin Oncol. 2013;10(8):472–84.

Gallegos MI, et al. Genetic heterogeneity in patients with multiple neoplastic lung lesions: a report of three cases. J Thor Oncol. 2007;2:12–21.

Hynes NE, et al. ERBB receptors and cancer: the complexity of targeted inhibitors. Nat Rev Cancer. 2005;5:341–54.

Jeuken JW, et al. MS-MLPA: an attractive alternative laboratory assay for robust, reliable, and semiquantitative detection of MGMT promoter hypermethylation in gliomas. Lab Invest. 2007;87:1055–65.

Lazar A, et al. Molecular diagnostics of sarcomas. Chromosomal translocations in sarcomas. Arch Pathol Lab Med. 2006;130:1199–207.

Metzker ML. Sequencing technologies—the next generation. Nat Rev Genet. 2010;11(1):31–46.

Michaloglou C, et al. BRAF(E600) in benign and malignant human tumours. Oncogene. 2008;27:877–95.

Moelans CB, et al. HER-2/neu amplification testing in breast cancer by multiplex ligation-dependent probe amplification in comparison with immunohistochemistry and in situ hybridization. Cell Oncol. 2009;31:1–10.

Nyati MK, et al. Nat Rev Cancer. 2006.

Rubin BP, et al. Gastrointestinal stromal tumour. Lancet. 2007;369:1731–41.

Sharma SV, et al. Epidermal growth factor receptor mutations in lung cancer. Nat Rev Cancer. 2007;7:169–81.

van Dongen JJ, et al. Design and standardization of PCR primers and protocols for detection of clonal immunoglobulin and T-cell receptor gene recombinations in suspect lymphoproliferations: report of the BIOMED-2 Concerted Action BMH4-CT98-3936. Leukemia. 2003;17:2257–317.

Zoll GJ, et al. General primer-mediated polymerase chain reaction for detection of enteroviruses: application for diagnostic routine and persistent infections. J Clin Microbiol. 1992;30:160–5.

Chapter 6
Molecular Diagnostics of Lymphoid Malignancies

A.W. Langerak and V.H.J. van der Velden

Abstract Blood cancer (leukaemia) and lymph node cancer (lymphoma) are increasingly being recognized as clinically heterogeneous diseases. It is therefore important to find markers with predictive value for disease course and survival. Such prognostic markers may also be used for treatment choices in individual patients. In addition, it is becoming increasingly important to sensitively and accurately monitor the effect of the chosen therapy and to adapt the therapy if necessary. This chapter highlights the role of molecular diagnostics in making a diagnosis, for prognostic classification, and for monitoring of lymphoid malignancies (leukaemias and lymphomas). In these assays, especially immunoglobulin (IG) and T-cell receptor (TCR) gene rearrangements are being used as molecular markers. Following an introduction on IG/TCR molecules and the basic mechanisms of the rearrangement processes in the IG/TCR gene complexes, strategies are discussed in which these DNA markers are used for molecular diagnostics of lymphoid malignancies. Next, a brief overview of other molecular markers and their applications in leukaemia and lymphoma diagnostics is given. Finally, new developments are discussed that in time will impact on molecular diagnostics of lymphoid malignancies.

Keywords Leukaemia · Lymphoma · Clonality · Prognosis · Monitoring · VDJ recombination · Immunoglobulin/IG · T-cell receptor (TCR) · MRD · Somatic hypermutation

---

A.W. Langerak (✉) · V.H.J. van der Velden
Laboratory for Medical Immunology, Department of Immunology,
Erasmus MC, University Medical Centre, Wytemaweg 80,
3015 CN Rotterdam, The Netherlands
e-mail: a.langerak@erasmusmc.nl

V.H.J. van der Velden
e-mail: v.h.j.vandervelden@erasmusmc.nl

E. van Pelt-Verkuil et al. (eds.), *Molecular Diagnostics*,
DOI 10.1007/978-981-10-4511-0_6

6.1 Introduction

6.1.1 IG/TCR Molecules and Their Gene Complexes

Human lymphocytes can recognize millions of different antigens in a very specific manner. This recognition is based on the enormous diversity (at least 10^{10}) of antigen-specific receptors (immunoglobulin [IG] and T-cell receptor [TCR] molecules) that are expressed on B- and T-lymphocytes, respectively. Each B- or T-lymphocyte expresses around 100,000 IG or TCR molecules with identical and unique antigen specificity. IG molecules consist of two IG heavy chains ("heavy chain"; IGH) connected by sulphur bridges, and two IG light chains ("kappa", IGκ or "lambda", IGλ). IG molecules are associated with CD79 molecules that mediate signals into the cell. TCR molecules consist of two chains that likewise are typically connected by sulphur bridges, and come in two variants: the "classic" TCRαβ receptor (combination of TCRα and TCRβ chain; 85–98% of peripheral blood T-lymphocytes) and the "alternative" TCRγδ receptor (combination of TCRγ and TCRδ chain; 2–15% of blood T-lymphocytes). Both types of TCR molecules are associated with CD3 molecules, which are necessary for signal transduction.

Each IG or TCR chain consists of at least two domains: a variable domain that is involved in the actual recognition of antigens and one or more constant domains that determine the effector function. The variable domains are encoded by a single exon that is formed by coupling of different DNA sequences. In this way, the required diversity of IG/TCR molecules can be created from a limited set of genes. In case of IGH, TCRβ and TCRδ chains (IGH, TCRB, TCRD gene complexes, respectively), this coupling concerns the combination of "variable" (V), "diversity" (D) and "joining" (J) genes, while the gene complexes encoding the IGκ, IGλ, TCRα and TCRγ chains (IGK, IGL, TCRA, TCRG gene complexes, respectively) only contain V and J genes (Table 6.1).

Table 6.1 Estimation of the potential primary repertoire of human IG/TCR molecules

	IG molecules			TCRαβ molecules		TCRγδ molecules	
	IGH	IGK	IGL	TCRA	TCRB	TCRG	TCRD
Number of functional genes[a]							
– V genes	40–46	34–37	30–33	45	44–47	6	6
– D genes	25	–	–	–	2[b]	–	3[b]
– J genes	6	5	4	50	13	5	4
Combinatorial diversity	$>2 \times 10^6$			$>2 \times 10^6$		>5000	
Junctional diversity	++	±	±	+	++	+	+++
Estimation total repertoire	$>10^{12}$			$>10^{12}$		$>10^{12}$	

[a]Numbers based on the IMGT (ImMunoGeneTics) international database (http://imgt.cines.fr/)
[b]In TCRD multiple D genes can be used in gene rearrangements; this means that the number of junctions can range from one to four. Gene rearrangements in TCRB typically contain only one D gene

6.1.2 V(D)J Recombination and IG/TCR Repertoire

During the development of precursor B- and T-cells in the bone marrow and thymus, respectively, the V, D and J genes of the various IG/TCR gene complexes are coupled through a strictly regulated process. This process is called gene rearrangement or V(D)J recombination (Fig. 6.1). The result is a lymphocyte-specific combination of V, D and J genes, also known as the V(D)J exon. During coupling of the V, (D) and J genes, deletion and random insertion of nucleotides occurs. This inaccurate coupling of the genes leads to an increase in diversity in the so-called "junctional regions" or "junctions" (Fig. 6.1).

The complete repertoire of IG/TCR molecules on B-and T-lymphocytes is determined by two levels of diversity: *combinatorial diversity* (different V(D)J combinations) and *junctional diversity* (due to the aforementioned inaccurate coupling of V, D and J genes).

Combinatorial diversity is the result of all possible combinations of available and functional V, D and J genes per IG/TCR locus, and of the pairing of two different functional protein chains per receptor molecule (IGH with IGκ or IGλ, TCRα with TCRβ and TCRγ with TCRδ). Since the human IGK, IGL, IGH, TCRA and TCRB loci consist of large numbers of V genes (30–45), sometimes D genes, and multiple J genes, a potential combinatorial diversity of $>2 \times 10^6$ IG/TCRαβ molecules can be achieved (Table 6.1). The combinatorial diversity of TCRγδ molecules is less, because of the limited number of functional V genes in these gene complexes.

Junctional diversity is based on the deletion of nucleotides at the ends of the genes involved in the rearrangement, as well as random insertion of nucleotides (N region nucleotides) between the coupled genes (junctional region). Insertion of N region nucleotides at the 3′ end of DNA breakpoints is mediated by the enzyme terminal deoxynucleotidyl transferase (TdT) in a template-independent manner. The junctional region of rearranged IG/TCR genes encodes for the complementarity determining region 3 (CDR3), which is involved in antigen recognition and which is actually a unique, lymphocyte-specific sequence ("fingerprint"). N region insertion therefore contributes to a drastic increase in the diversity of antigen recognition by IG/TCR chains and molecules. This is particularly true for IGH, TRB and especially TRD gene rearrangements in which the junctional region consists of multiple couplings (V-D, D-J and even D-D) (Table 6.1).

Although the estimates on the number of different IG/TCR molecules assume random use of available functional V, D and J genes, there is evidence for preferential use of these genes. $TCR\alpha\beta^+$ T-lymphocytes for example preferentially use the Jβ2 genes over the Jβ1 genes, while peripheral $TCR\gamma\delta^+$ T-lymphocytes preferentially show Vγ9-Jγ1.2 and Vδ2-Jδ1 rearrangements. An alternative and more likely explanation is that gene usage as such is stochastic, and that the over-representation of certain receptor types is due to antigenic selection and expansion of cells with certain receptors in peripheral tissues.

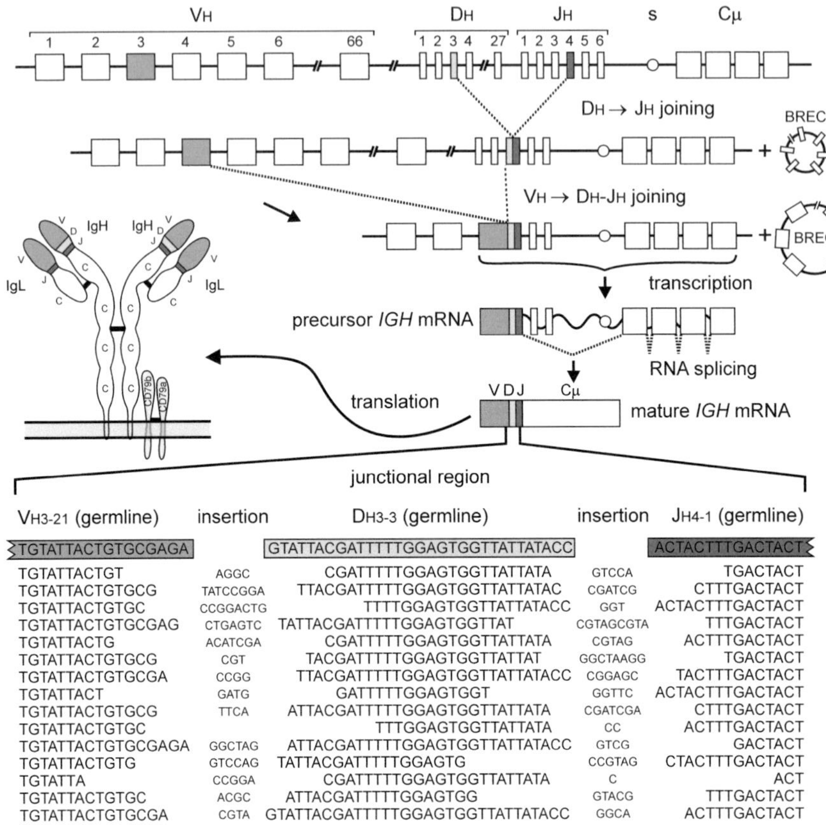

Fig. 6.1 Recombination and diversity in the human IGH locus. The IGH gene complex contains ~65 VH genes, 27 DH genes and 6 JH genes. In this example the first rearrangement concerns the DH3 and JH4 genes, and is followed by a second recombination of the VH3 gene to the DH3-JH4 combination, resulting in the formation of a VH4-DH3-JH4 coding joint. During the recombination process also two extra-chromosomal B-cell receptor excision circles (BRECs) are formed, which respectively contain the D-J and V-D signal joints. As such excision circles do not replicate, they are slowly diluted during proliferation of the (precursor) lymphocytes. Extra diversity in the coding joints is caused by deletion (at the V, D, and J ends) and random insertion of nucleotides. In this example, all coding joints represent the same V(D)J combination, but all are unique in their coupling areas (junctional regions); this is known as junctional diversity. Uppercase and lowercase letters indicates DH genes and inserted nucleotides, respectively. The rearranged IGH gene is transcribed into precursor mRNA, eventually spliced into mature mRNA, and finally translated into a IgM chain. Because V(D)J recombination is a complex process with inaccurate coupling of genes, roughly two out of three sequences will not contain the correct reading frame. This high frequency of out-of-frame rearrangements explains why most B- and T-lymphocytes show rearranged IG or TCR genes on both alleles

6.1.3 Leukaemias and Lymphomas as the Malignant Counterparts of Precursor (Lymphocytes)

The various acute and chronic and lymphomas of B- and T-cell origin can be considered as the malignant counterparts of (precursor) B- and T-lymphocytes (Fig. 6.2). In addition to morphological and cytochemical characterization, normal (precursor) B- and T-lymphocytes and their malignant counterparts can also be compared by means of immunophenotyping with antibodies directed against (membrane-bound or intracellular) antigens. Because IG/TCR gene rearrangements occur from the earliest stages of lymphoid differentiation, they are present in virtually every immature and mature lymphocyte. Leukaemias and lymphomas, which arise from malignantly transformed (precursor) lymphocytes, thus contain one or more identical (=clonal) IG/TCR gene rearrangements. This forms the basis for using rearranged IG/TCR genes as molecular markers in molecular diagnostics of leukaemias and lymphomas.

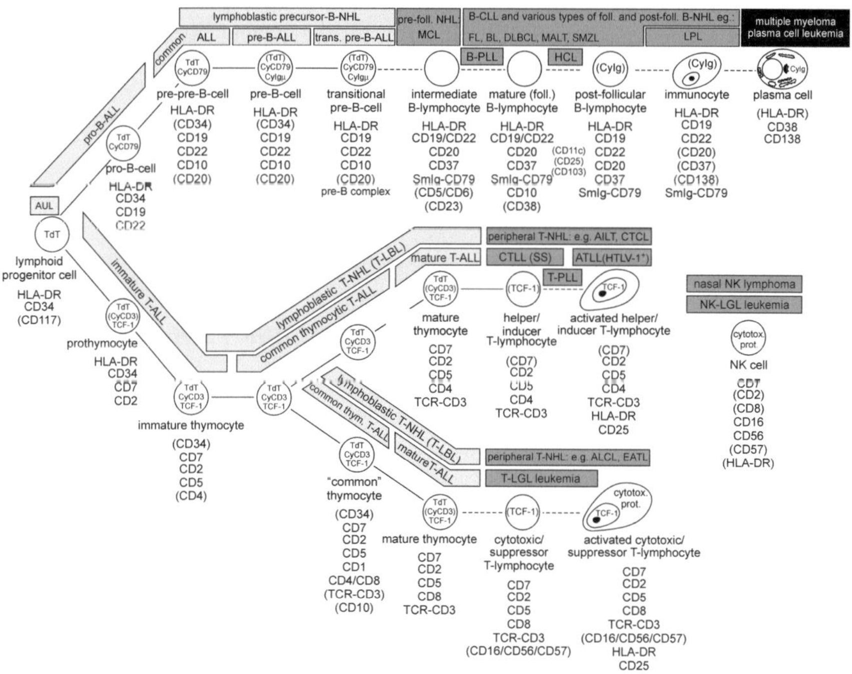

Fig. 6.2 Leukaemias and lymphomas are the malignant counterparts of normal lymphoid cells in various stages of development. The bars show the different types of acute leukaemias (*light grey*), mature leukaemias and non-Hodgkin's lymphomas (NHL) (*dark grey*) and multiple myelomas (*black*), ordered by the suspected cell of origin. For the various cells that can be distinguished during development the relevant immunophenotypic markers are indicated; markers in parentheses are not always detectable. For abbreviations see Table 6.2

6.2 Diagnosis: IG/TCR Clonality Testing

Tumour cells are the offspring of a single malignantly transformed cell. Hence all tumour cells are clonally related. This monoclonality is the major characteristic of malignant tumour cell populations, and differentiates them from oligoclonal or polyclonal, reactive cell populations. In leukaemias and lymphomas, clonality testing is relatively easy, due to the widespread availability of highly polymorphic DNA markers, that is IG/TCR gene rearrangements (see above). Heterogeneity of IG/TCR gene rearrangements correlates with polyclonally activated cells, while identically rearranged IG/TCR genes are compatible with a monoclonal lymphoid cell population.

Prior to the PCR-era, IG/TCR clonality testing was mostly carried out via Southern blot analysis. Although time consuming and laborious, the Southern blot method has always been reliable, with a high specificity and fair sensitivity (detection limit: size of the malignant clone $\geq 10\%$ of the total cell population). A very significant drawback, however, is that relatively large amounts of high-molecular DNA are required, which hampers routine analysis of small biopsies and/or paraffin fixed material. For this reason, the PCR technology is a good alternative, with several PCR strategies for IG/TCR clonality testing having been developed and described. Almost all of these PCR strategies, however, suffer from two main pitfalls:

1. False negativity, by poor primer annealing to some genes
2. False-positivity, when insufficient distinction between polyclonal and monoclonal gene IG/TCR rearrangements is made.

6.2.1 *Standardization of PCR-based Clonality Testing*

At the end of the nineties a European consortium of ~45 laboratories (BIOMED-2 Concerted Action BMH4-CT98-3936) aimed to develop a reliable standard in the field of PCR-based clonality testing (van Dongen et al. 2003). The problem of false negativity was approached in several ways:

1. Developing complete primer sets that recognize all possible V-J rearrangements of the IG/TCR loci
2. Inclusion of additional targets in the form of incomplete rearrangements (e.g. DH-JH and Dβ-Jβ)
3. Analysis of multiple IG targets (IGK and IGL next to IGH) and multiple TCR targets (TCRB and TCRD besides TCRG).

This strategy of complementary and complete recognition of targets could in practice only be achieved through the development of so-called multiplex PCR reaction mixes that consist of multiple forward and/or reverse primers. As the

IG/TCR loci consist of many V and J genes that can rearrange, many PCR reactions of one forward and one reverse primer would be needed to amplify all possible combinations of V and J genes in the various IG/TCR loci. In the most extreme scenario of the TCRB locus, this would imply 23 × 13 reactions (23 forward primers for the 23 V families and 13 reverse primers for the 13 J genes). By careful design with software and testing of multiplex primer combinations in the European consortium, it has been possible to limit the number of PCR reactions that are required to evaluate all possible IG/TCR rearrangements. In the case of TCRB, the 23 × 13 = 299 reactions were replaced by only two multiplex reactions. The other challenge was to avoid false-positivity. This was achieved by introducing standardized, reliable methods for evaluation of PCR products: heteroduplex analysis and in particular Gene Scan fragment analysis (Fig. 6.3). While both methods can be used to evaluate the heterogeneity of PCR products, there are intrinsic differences. High-resolution Gene Scan fragment analysis evaluates size differences of the PCR products, whereas heteroduplex analysis allows the user to assess variations in size and composition of the nucleotides in the junctional regions of rearranged IG/TCR genes. In heteroduplex analysis, monoclonal IG/TCR PCR products, as they are found in leukaemias and lymphomas, result in a clear band when analysed with gel electrophoresis, while a smear of heterogeneous PCR products is seen for non-leukemic proliferations. In Gene Scan analysis, monoclonal PCR products give rise to a dominant peak of fluorescent products of identical length, whereas polyclonal PCR products derived from a reactive lymphoid lesion result in a Gaussian distribution profile (Fig. 6.3).

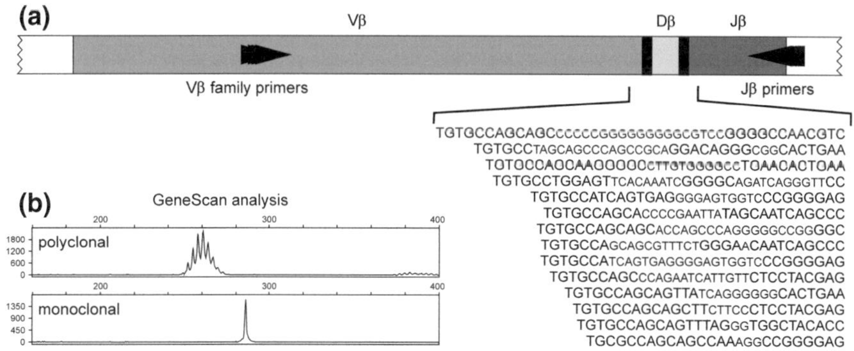

Fig. 6.3 IG/TCR clonality testing by means of Gene Scan fragment analysis. **a** Rearranged IG/TCR genes (in this example TCRB rearrangements) showing heterogeneous junctional regions (CDR3 motifs) that differ in size and composition of nucleotides. The V, D, and J gene germline nucleotides are shown in capital letters and the randomly inserted nucleotides in small letters. The diversity of the junctional regions can be used in Gene Scan fragment analysis to distinguish between polyclonal (heterogeneous) and monoclonal (identical) PCR products. **b** In Gene Scan fragment analysis denatured fluorochrome-labelled PCR products are analysed by high-resolution fragment analysis. Monoclonal PCR products of identical size, as in the case of T-cell leukaemia, show a clear peak, while polyclonal PCR products of, for example, a reactive lymphocyte population show a Gaussian distribution (normal distribution) of product sizes

Table 6.2 Results of PCR-based IG/TCR clonality testing in lymphoid malignancies

	n	IGH (%)	IGK (%)	IGH + IGK (%)	IGL (%)
MCL	54	100	100	100	44
CLL/SLL	56	100	100	100	30
FL	109	86	84	100	21
MZL[a]	41	95	83	100	29
DLBCL	109	85	80	98	28
	n	TCRB (%)	TCRG (%)	TCRB + TCRG (%)	TCRD (%)
T-PLL	33	100	94	100	6
T-LGL	28	96	96	100	29
PTCL-NOS	47	98	94	100	15
AITL	37	89	92	95	35
ALCL[b]	43	74[b]	74[b]	79[b]	9[b]

Abbreviations AITL Angioimmunoblastic T-cell lymphoma; *ALCL* Anaplastic large cell lymphoma; *CLL* Chronic lymphocytic leukaemia; *DLBCL* Diffuse large B-cell lymphoma; *FL* Follicular lymphoma; *MCL* Mantle cell lymphoma; *MZL* Marginal zone B-cell lymphoma; *PTCL-NOS* Peripheral T-cell lymphoma not otherwise specified; *T-LGL* T-cell large granular lymphocyte leukaemia; *T-PLL* T-cell prolymphocytic leukaemia

[a]MZL includes both the extranodal (31) as well as the nodal (10) type

[b]Lower rates in ALCL are partly the result of a series of nine null-type ALCL without any TCR rearrangement

Following a technical evaluation phase, the multiplex IG/TCR PCR protocol was validated using ~600 lymphomas, defined according to WHO criteria (van Krieken et al. 2007). Very high frequencies of clonal rearrangements were shown in all lymphoma entities. In B-cell malignancies, high frequencies of IGH clonality were found (Table 6.2), while a further increase in clonality detection was seen when IGK was added as a target (especially in leukaemias/lymphomas of (post-) germinal centre origin). IGL analysis appeared to have little added value. The complementary value of the analysis of multiple IG targets in (post-)germinal centre B-cell leukaemias and lymphomas is related to poor primer adhesion, which is due the presence of somatic hyper mutations (SHM) in certain IG genes (see Sect. 6.3.1 for more details on germinal centre response and SHM). In the various T-cell malignancies, a high frequency of clonality was seen by combined TCRB and TCRG analysis. TCRD analysis in isolation showed little added value (Table 6.2). The BIOMED-2 multiplex PCR strategy as discussed here is now the worldwide standard.

6.2.2 *Pitfalls and Interpretation*

With the technical aspects being standardized, correct interpretation of the IG/TCR clonality testing results becomes very important (Langerak et al. 2007, 2012). Knowledge of the biological processes around IG/TCR recombination and experience with clonality testing is essential in that respect. Awareness of various immunobiological pitfalls is crucial to avoid misinterpretation of the results. These concern, amongst others, pseudoclonality (co-incidental, non-random amplification) due to small numbers of T or B lymphocytes, oligoclonality, which occurs for example in an aging immune system, or (mono)-clonality in reactive lesions with a dominant repertoire of T or B lymphocytes. In addition, technical pitfalls, such as poor quality DNA, should be taken into consideration as well. Finally, optimal integration of histomorphological data (and preferably also flow cytometry data) is an absolute requirement for correct interpretation of molecular data. Direct interaction between relevant specialists (molecular biologists, immunologists, haematologists and pathologists) is the best guarantee for the integration of all available data streams into the most reliable diagnosis. To ensure high quality clonality testing, regular quality assurance rounds are being organized by the international EuroClonality network.

6.2.3 *Applications*

IG/TCR clonality testing has multiple applications (Langerak et al. 2007). The distinction between tumour cells on the one hand and normal reactive lymphocytes on the other hand, is the most important application. Clonality testing contributes to making a timely and accurate diagnosis of leukaemia or lymphoma in suspicious lesions. IG/TCR clonality testing is also very useful to establish the clonal relationship between multiple lymphoproliferations. Such lesions can occur at the same time, or over time, in case of a return of the disease (recurrence or new malignancy). In such situations, a direct and accurate comparison of samples in the same experiment is essential. Evaluation of metastases/disease spread and monitoring of therapy effectiveness can also be done via clonality testing, although these are limited by the analytical sensitivity of the assays (1–10% depending on the actual IG/TCR target). A more sensitive approach for measuring metastases or for therapy monitoring requires real-time quantitative PCR-based strategies with primers and patient-specific probes (see Sect. 6.4).

6.3 Prognostic Classifcation: IGHV Somatic Hyper Mutation (SHM) Analysis

In certain situations, a diagnosing of leukaemia or lymphoma per se is not sufficient, because the prognosis of a certain type of leukaemia/lymphoma may vary between individuals. Some leukaemia or lymphoma types are clinically heterogeneous and exhibit differences in survival. For example, chronic lymphocytic leukaemia (CLL) of the B-cell type typically presents as an indolent disease, although in some patients the disease is more aggressive. In the last years, several clinical and biological parameters have been associated with differences in survival of CLL patients.

Amongst these, the IGHV somatic hyper mutation (SHM) status shows a very clear prognostic value. CLL with a rearranged IGH gene, but without somatic mutations in the IGHV gene ("unmutated CLL") typically have a less favourable prognosis than CLL with mutations in the IGHV gene ("mutated CLL") (Damle et al. 1999; Hamblin et al. 1999).

6.3.1 *SHM as a Physiological Phenomenon*

SHM is a molecular process that occurs during a physiological immune response. Upon encountering an antigen, naive B lymphocytes in the germinal centre, a specialized micro-environment in the lymph node, undergo SHM in the V(D)J exons encoding the IG heavy and light chain variable domains (Fig. 6.4). Mutations

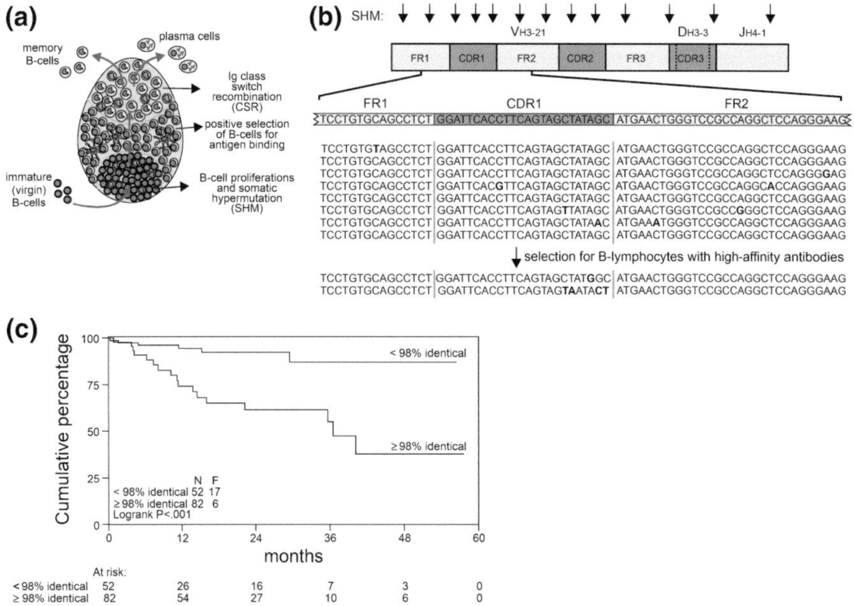

Fig. 6.4 IGHV somatic hypermutation (SHM) status as the basis for determining prognosis in CLL. **a** Schematic representation of a germinal centre of a lymph node indicating the most relevant cell types and regions involved in IG somatic hyper mutation and IG class switch recombination. **b** During the SHM process mutations are randomly introduced in the rearranged IG gene complexes, resulting in additional diversity as shown here for the VH3-21-DH3-3-JH4 rearrangement. From the pool of B lymphocytes with unique, mutant sequences, only some are selected on the basis of high affinity antibodies. **c** Survival analysis (in months) of CLL on basis of IGHV SHM status shows the better prognosis for CLL with mutated IGHV genes (<98% identical) compared to CLL with unmutated IGHV genes ($\geq$98% identical)

can occur in a large area of the V(D)J exon, with the highest frequencies in the V genes of the IG loci. Although the SHM process is in principle random, it will eventually lead to affinity maturation through selection of mutated IG molecules. Following SHM the B-lymphocytes with the best fitting IG receptors are selected based on their affinity for antigens that are presented to the B-lymphocytes in the germinal centre (germinal centre reaction). This leads to the formation of memory B-lymphocytes that may eventually differentiate into antibody-producing plasma cells. On the basis of several studies currently two types of CLL are recognized, namely mutated CLL arising from memory B cells that have undergone a germinal centre response and unmutated CLL that probably originated from a separate pool of activated B-lymphocytes without germinal centre response. The difference in SHM status between the two types of CLL thus reflects their different biological origin and appears to be associated with differences in survival.

6.3.2 *Technical Aspects of IGH SHM Status Evaluation*

In order to determine the IGHV SHM status in CLL patients genomic DNA or RNA/cDNA can be used as a starting material (Ghia et al. 2007). In case of genomic DNA, multiplex PCR reactions are performed with IGH leader or IGHV family primers in combination with an IGHJ consensus primer; when cDNA is the starting material, IGH leader or IGHV family primers are combined with an IGHC primer (directed against the C exon, which is only connected to the VDJ exon at the transcript level). Following amplification, additional analysis is needed to determine whether a monoclonal PCR product is formed; if multiple clonal PCR products are present separate PCR reactions are performed with one IGH leader primer or IGHV primer. The PCR products are first sequenced from the IGHJ or IGHC side. Next, a reaction from the IGHV side is used to confirm the obtained sequence. In this way a consensus sequence is obtained. Via web-based programs, such as the IMmunoGeneTics (IMGT) (http://imgt.com/) or IgBlast (http://www.ncbi.nlm.nih.gov/igblast/) databases, the involved V, (D) and J genes are identified. IMGT is the most appropriate database as it is most updated with respect to polymorphic variants of IGHV genes in the human population. Apart from identification of the involved genes, IMGT can also be used to determine the IGHV SHM status of an IGH PCR product. According to the guidelines of the European Research Initiative on CLL (ERIC; http://www.ericll.org/), only the IGH SHM status of in-frame, productive rearrangements (i.e. without stop codons in the CDR3 junctional region) can be reported (Ghia et al. 2007; Langerak et al. 2011). The IGHV SHM status is calculated as the percentage of identical IGHV nucleotides compared with the most similar germline IGHV gene in the genome. Based on this percentage CLL are classified as mutated (<98% identical) or unmutated ($\geq$98% identical). The 98% cut-off was originally chosen to prevent the risk that (unknown) polymorphic variants would be counted as mutations; however, in the meantime this cut-off has proven its clinical relevance in several clinical studies (Fig. 6.4c).

◂**Fig. 6.5** Detection of minimal residual disease. **a** Schematic representation of the kinetics of tumour cells during and after therapy for ALL. The detection limit of cytomorphological techniques is between 1 and 5% and thus provides only limited insight into the presence of remaining tumour cells. By using more sensitive techniques (such as immunophenotyping and PCR-based analysis) also lower numbers of remaining tumour cells can still be detected (1 tumour cell in 100,000 normal cells). This allows a much better view on the behaviour of the tumour cells during and after therapy. **b** Relapse-free survival of children with ALL treated according to the ALL8/BFM90 protocol. The children are divided into three risk groups based on MRD levels on day 33 and day 79 after start of treatment. Patients in the low-risk group have undetectable MRD levels at both time points, high-risk patients show MRD levels of 10^{-3} or higher at both times, and the remaining patients form the intermediate-risk group. Numbers of patients in the different risk groups are in parentheses. **c** Event-free survival (EFS) of children with ALL who underwent stem cell transplantation, classified based on the MRD level before transplantation. The 5-year EFS and the number of patients per group is shown at the end of each curve

6.3.3 IGH Gene Usage, CDR3 Junctions and Stereotyped IG Receptors

Extensive studies in large CLL cohorts have shown that certain IGHV genes are preferentially associated with unmutated or mutated CLL. The IGHV1-69 gene is often seen in the unmutated subset (with the less favourable prognosis), while for example the IGHV4-34 and IGHV3-21 genes are more frequently seen in mutated CLL. These associations most probably reflect involvement of specific antigens in activation of the memory-type B-lymphocytes from which the CLL arise (see Sect. 6.3.2). A second argument for the important role of antigens in CLL is the occurrence of so-called stereotyped IG receptors that are characterized by highly similar to (virtually) identical CDR3 junctions. Recognition of CLL subsets with stereotyped receptors not only has pathobiological implications, but also shows clinical prognostic value. The best example of this is the mutated stereotyped IGHV3-21 CLL subset with characteristic CDR3 junctions, which independent of the IGHV SHM status points to poor prognosis (Baliakas et al. 2014).

6.3.4 Applications

The IGHV SHM status in CLL patients is one of the most important prognostic factors. The IGH SHM results can be used to determine eligibility of patients for particular clinical trials. It is therefore important that the implementation and, in particular, the interpretation of the IGHV SHM status test are standardized, because only then can a direct comparison of IGHV data in multi-centre trials be possible. An important note is that the prognostic significance of the IGHV SHM status for the individual CLL patient is limited, since IGHV SHM status-based survival data are defined for groups of patients. Rather, for individual CLL patients, a prognostic risk profile, not only taking into account IGHV SHM status, but also other known prognostic factors, is probably of greater significance.

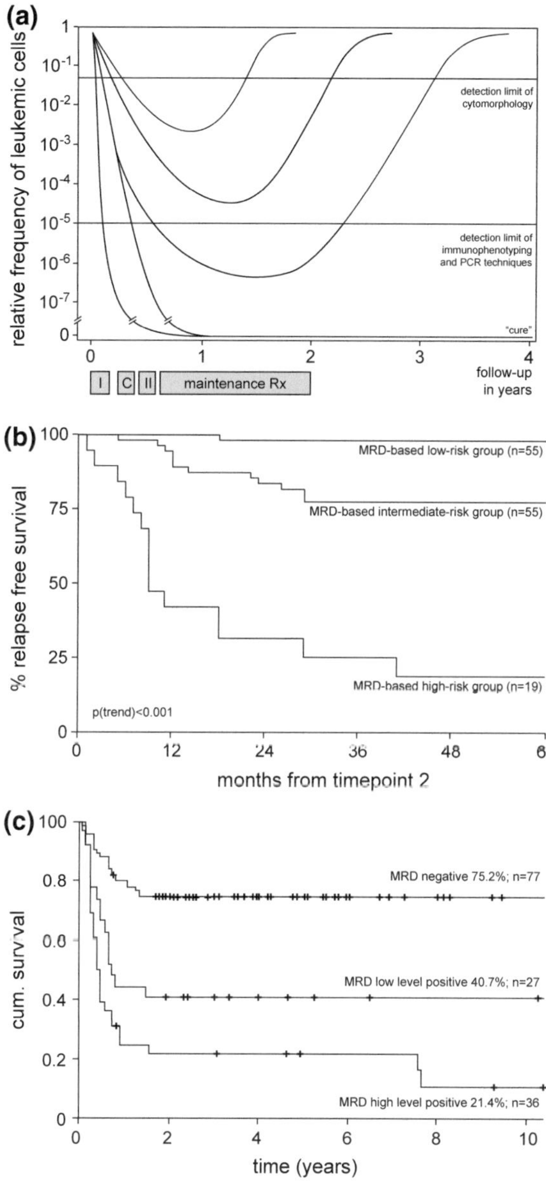

6.4 Monitoring: Detection of Minimal Residual Disease by IG/TCR Targets

Modern treatment protocols induce complete remission in the majority of patients with lymphoid malignancies. Nevertheless, relapses (disease recurrence) occur in a significant proportion of patients.

Apparently, the treatment protocols are not capable of killing all clonogenic malignant cells in these patients, even though they reached complete remission according to cytomorphological criteria. The detection limit of cytomorphological techniques is however not lower than 1–5% of malignant cells, implying that these techniques can provide only superficial information about the effectiveness of the treatment. More sensitive techniques are required for the detection of low frequencies of malignant cells during and after treatment, i.e. for the detection of minimal residual disease (MRD) (Fig. 6.5).

Detection of MRD is mainly performed through PCR analysis of IG/TCR gene rearrangements (DNA level), because this method can be used in the vast majority of patients with lymphoid malignancies. Moreover, this method generally results in very good sensitivities of 0.01% (10^{-4}, one tumour cell in 10,000 normal cells). Since it is not only important to detect the presence or absence of malignant cells, but also to determine the amount of residual malignant cells, qPCR analysis is generally used. MRD analysis using qPCR analysis of IG/TCR gene rearrangements involves four steps:

1. Identification of MRD-PCR targets in the diagnostic sample
2. Design of a target-specific qPCR assay
3. Sensitivity testing of the qPCR using serial dilutions of the diagnostic sample
4. MRD analysis of follow-up samples and interpretation of the data.

6.4.1 Detection and Identification of IG/TCR Gene Rearrangements

During early B-and T-cell development, rearrangements of the IG/TCR gene complexes take place. At the joining sites of the V-D and D-J genes, random deletions and insertions of nucleotides occur, resulting in a cell-specific "junctional region". These junctional regions are different in each normal lymphocyte; however, since all tumour cells will be derived from a single normal cell, the junctional region of all malignant cells will be identical. This junctional region can therefore be considered as a patient-specific "fingerprint" of the tumour cells. In >95% of the patients with a lymphoid malignancy, the junctional regions of the rearranged IG/TCR genes can be identified at diagnosis (by PCR Gene analysis or by PCR heteroduplex analysis followed by sequencing) (see Sects. 6.2 and 6.3) (Pongers-Willemse et al. 1999).

6.4.2 Design of a MRD Target-Specific qPCR

By designing specific oligonucleotides that can bind to the junctional region of the tumour cell, the malignant cells can be detected during and after therapy. The junctional region-specific oligonucleotide, also referred to as the allele-specific

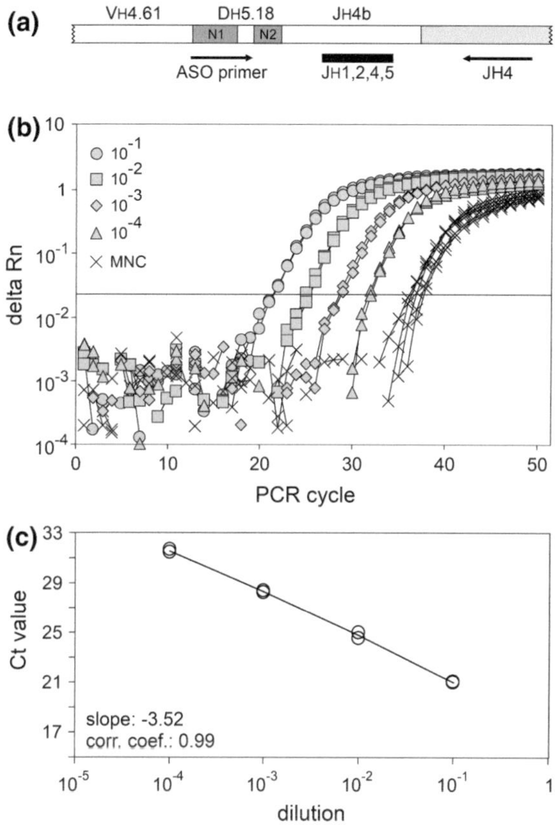

Fig. 6.6 qPCR analysis of an IGH gene rearrangement. **a** Schematic representation of a IGH gene rearrangement. The patient-specific (ASO) primer is positioned in such a way that it specifically binds to the junctional region of the IGH gene rearrangement from the ALL patient. This primer is used in combination with a germline probe (JH1, 2, 4, 5) and a germline reverse primer (JH4). **b** qPCR amplification plots of the patient-specific IGH gene rearrangement. To determine the sensitivity and specificity of the patient-specific primer, tenfold dilutions are made from DNA at diagnosis into DNA of normal mononuclear cells (MNC); in this example a sensitivity of 10^{-4} is obtained. **c** By plotting the C_q value (the PCR cycle in which the fluorescence signal [Rn] comes above the threshold, also called C_T) against the dilution, a calibration curve can be obtained. C_q values obtained by qPCR analysis of samples during and after treatment can in this way be converted to MRD levels

oligonucleotide (ASO), is usually a primer, but ASO probes can be used as well (van der Velden et al. 2003). In the so-called ASO forward primer approach (Fig. 6.6a), the forward primer (in particular the 3′ end) is positioned in or around the junctional region and is used in combination with a germline probe and a germline reverse primer (for example located in the J gene). The ASO reverse primer approach is similar, but with opposite positions of the germline probe and primer to the junctional region (so often localized in the V gene). For the qPCR-based detection of tumour-specific IGH, IGK, IGL, TCRG, TCRD, and TCRB gene rearrangements, several primer-probe sets have been developed. The ASO forward primer approach has a number of advantages over the ASO reverse primer approach. First, the number of J genes is significantly lower than the number of V genes, so that much less (relatively expensive) germline probes are needed. Secondly, the ASO forward primer approach is less sensitive to false negativity, because, for example, the DH-JH junction will be maintained in ongoing or secondary VH to DH-JH rearrangements and in IGHV replacements (see Sect. 6.4.5). Thirdly, the occurrence of somatic hyper mutations in the V gene may result in less optimal binding of the germline probe and the germline reverse primer. On the other hand, the ASO reverse primer approach can be more sensitive in some cases, because there are many more V- than J genes, making the qPCR more specific and thereby more sensitive.

6.4.3 qPCR Sensitivity Testing

The required sensitivity of the qPCR assay depends on the clinical application, but in general should reach a sensitivity of at least 10^{-3}, preferably 10^{-4} to 10^{-5}. To determine the sensitivity of the qPCR assay, the diagnostic material (containing the patient-specific junctional region) is serially diluted in DNA obtained from healthy donors (containing polyclonal junctional regions) and analysed in the qPCR (Fig. 6.6b). Generally, serial dilutions are made ranging from 10^{-1} (10% tumour cells) down to 10^{-6} (0.0001% tumour cells). The DNA of healthy donors must also be analysed separately (in at least six wells) to determine the specificity of the assay. Ideally, the polyclonal junctional regions in the healthy controls will not show any amplification by the applied qPCR. However, if gene rearrangements with comparable junctions to the tumour cell are present, some non-specific amplification may be observed, since the ASO primer may also bind to these normal junctional regions. By plotting the logarithmic value of the serial dilutions (i.e. the percentage of tumour cells) against the "quantification cycle" (C_q), a standard curve can be obtained. This standard curve should have a slope of about −3.3 (Fig. 6.6c) (van der Velden et al. 2003).

Several factors must be taken into account when determining the sensitivity of the qPCR assay. These include the reproducibility of the measurement (difference in Cq values of duplicates/triplicates), the difference between the specific and non-specific amplification, and the slope and correlation coefficient of the standard curve. It is particularly important that the specific amplification is sufficiently apart from the non-specific (background) amplification, so that false-positive or false-negative results can be prevented (see also Volume 1, Sect. 5.2.4). In order to compare the sensitivities of the qPCRs between different laboratories and between different clinical studies, it is essential to have uniform definitions for the sensitivity. For IG/TCR-based qPCR, data in ALL guidelines have been developed within the international EuroMRD network (previously the "European Study Group on detection in ALL; ESG-MRD-ALL), a consortium of approximately 45 international laboratories (van der Velden et al. 2007a, b).

The sensitivity of an MRD-PCR analysis with a junctional region as target depends on the type of rearrangement, the size and composition of the junctional region, and the "background" of normal lymphoid cells with a similar IG/TCR gene rearrangement. The junctional regions of complete V-D-J rearrangements (IGH, TCRB, TCRD) are large, while the junctional regions of V-J rearrangements (IGK, IGL, TCRG) are generally three to four times smaller. Consequently, the sensitivity of a qPCR using the latter rearrangements is often lower than the sensitivity of the V-D-J rearrangements. In addition, the sensitivity of the qPCR may be dependent on the sampling time point of the follow-up sample. In heavily regenerating bone marrow (as seen after stopping of (chemo)therapy), large numbers of normal precursor B-cells with polyclonal IGH gene rearrangements may be present. These normal IGH gene rearrangements may result in an increased background (non-specific amplification) of the qPCR and thereby in reduced sensitivity (Fig. 6.7).

6.4.4 qPCR Analysis of Follow-up Samples

To analyse the MRD level in a follow-up sample, the patient-specific qPCR analysis is performed on the follow-up DNA samples together with the serial dilutions of the diagnostic sample and DNA of healthy donors. The C_q values of the follow-up samples can be plotted into the standard curve, so that the corresponding MRD level can be calculated.

To be able to quantify the MRD data, it is necessary to check the quality and quantity of the DNA by running a reference gene qPCR (often the albumin gene is used). For reliable quantification, it is necessary that the exact same DNA sample is used for the IG/TCR qPCR and the reference gene qPCR.

Although the qPCR analysis is in principle a quantitative approach, it is not always possible to quantify the data precisely. Especially low levels of MRD, below the reproducible range of the assay, cannot be reliably quantified. Also, it can be difficult to make a clear distinction between MRD-positivity and MRD-negativity; this is especially true for MRD-PCR targets for which non-specific amplification is observed, since background levels may vary between different follow-up time points (Fig. 6.7).

Sometimes the results of the reference gene qPCR are lower than expected. These lower values can be the result of inhibition, which can be seen in a substantial number of bone marrow and blood samples (5–10%). The addition of 0.04% BSA prevents such inhibition and it is therefore recommended to routinely add 0.04% BSA to the qPCR reaction. If a sample unexpectedly does not meet predefined criteria for the reference gene values, one has to be extremely careful with the interpretation of the data, particularly if the sample seems to be MRD negative. In this case, the results may not be acceptable for further evaluations or it should be indicated clearly that the DNA quantity/quality of the sample was below expectations.

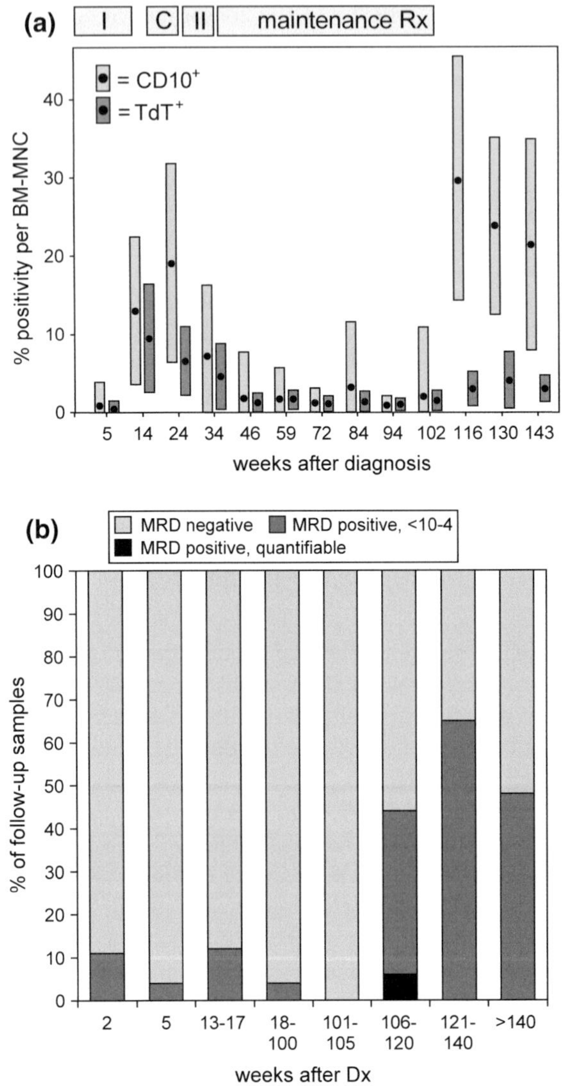

Fig. 6.7 B-cell regeneration and non-specific amplification (NSA) of bone marrow samples obtained during and after therapy. **a** Frequency of CD10+TdT+ and CD10+TdT- B-cell precursors in bone marrow samples taken from childhood ALL patients at several time points during and after therapy according to the DCOG ALL-8 protocol. Data are expressed as median and 50% confidence interval. **b** RQ-PCR data interpretation. Five patient-specific IGH targets, showing some NSA in normal buffy coat DNA, were tested by qPCR on bone marrow samples of five other ('non-related') ALL patients that were MRD negative (mean samples per patient: 12). Data were interpreted according to EuroMRD guidelines and expressed as 'MRD-negative', 'MRD-positive, $<10^{-4}$', or a quantitative MRD result ('MRD-positive, quantifiable'). Although the follow-up samples are MRD negative, some amplification can be observed, particularly in those follow-up samples with high numbers of normal B-cells

To compare MRD data from various studies and/or laboratories with each other, it is crucial to apply uniform guidelines for performance and interpretation of the qPCR. For IG/TCR-based MRD analysis in ALL, such guidelines have been developed within EuroMRD (van der Velden et al. 2007a). The same guidelines are nowadays also being used for analysis in mature lymphoid malignancies, such as B-NHL.

6.4.5 *Potential Pitfalls When Using IG/TCR Gene Rearrangements as MRD-PCR Target*

qPCR analysis of rearranged IG/TCR genes is hampered by three phenomena that can result in false-negative MRD results: oligoclonality, secondary and ongoing rearrangements, and somatic hyper mutations.

6.4.5.1 Oligoclonality

IG/TCR gene rearrangements in hematologic malignancies (especially in ALL) can be subject to sub-clone formation. The problem of oligoclonality at diagnosis is that it is unknown which clone may eventually result in a relapse and therefore should be monitored with the MRD-PCR assay. Several studies in ALL have shown that monoclonal MRD-PCR targets are generally very stable, while oligoclonal MRD-PCR targets are often lost at relapse. Despite these immunogenotypic changes between diagnosis and relapse, in 75–90% of the precursor-B-ALL patients and in 90% of T-ALL patients at least one rearranged IGH, TCRG and/or TCRD allele remains stable. In most ALL patients, at least two MRD-PCR targets with sufficient sensitivity in the qPCR are available. It is therefore generally accepted that for MRD-PCR studies in ALL, at least two IG/TCR targets per patient should be used, drastically reducing the likelihood of obtaining false-negative MRD results. In chronic leukaemias and lymphomas oligoclonality is rare.

6.4.5.2 Secondary and Ongoing Rearrangements

In the time period between diagnosis and disease recurrence, secondary IG/TCR gene rearrangements (replacing the initial DH-JH VH-JH, or Vk-Jk rearrangements by rearrangements that use more upstream located V genes and more downstream located J genes) and ongoing IG/TCR gene rearrangements (for example, rearrangement of a VH to a DH-JH rearrangement or IGHV "replacements") result in the loss of the tumour-specific MRD-PCR targets. Because the DH-JH junction remains preserved in V to D-J rearrangements and in V replacements, it is recommended to design the primers around this relatively stable D-J junction, thereby reducing the chance of obtaining false-negative MRD results as much as possible. Secondary gene rearrangements hardy occur in chronic leukaemias and lymphomas.

6.4.5.3 Somatic Hypermutations

If rearranged IG genes are used as MRD-PCR targets, one should be aware that somatic hyper mutations (SHM) could affect the binding of the primers used. The degree of SHM is relatively high in multiple myeloma (median of 8% mutated nucleotides) and relatively limited in CLL (2%) and FL (4%). Because IGK-Kde and DH-JH rearrangements are not subject to SHM, these rearrangements are particularly suitable for MRD analysis in chronic leukaemias and lymphomas.

6.4.6 Clinical Applications of MRD Detection

Detection of MRD through PCR analysis of IG/TCR gene rearrangements is possible in virtually every patient with a lymphoid malignancy. The clinical application, however, depends on the type of malignancy (Table 6.3).

6.4.6.1 Clinical Significance of MRD Detection in ALL

The most relevant application of MRD in children and adults with ALL is the evaluation of the initial treatment response. Several studies have shown that MRD is of strong and independent prognostic significance. Based on MRD levels determined in bone marrow samples obtained at two time points early after start of treatment, three patient groups with significantly different clinical outcome can be

Table 6.3 Prognostic significance and clinical application of MRD detection in lymphoid malignancies

Disease category	Type of MRD application			
	Early response to therapy	Continuous for treatment adaptation	MRD analysis pre-HSCT	MRD analysis post-HSCT
ALL	+++	++	+++	++
CLL	+++	+*	+	++
B-NHL	–	+*	+	++
Chronic T-cell leukaemia	–	+*	–	–
T-NHL	–	+*	–	–

+++: Importance of MRD detection proven in large prospective studies
++: Probably clinically relevant (for example, in a subset of patients) but not yet proven in large prospective studies
+: MRD results are statistically significant but clinical implementation is still limited
*: Only relevant for patients treated with intensive therapy and/or CD20 antibody therapy
–: MRD detection has no added value in comparison with cytomorphology

recognized (Fig. 6.5b) (van Dongen et al. 1998). In current treatment protocols for ALL, MRD-based high-risk patients receive intensified therapy (including stem cell transplantation in first complete remission or experimental therapies) (Flohr et al. 2008). In contrast, MRD-based low risk patients (about 30–40% of all ALL patients, which is comparable to the percentage of patients surviving before the introduction of intense chemotherapy) currently receive significant therapy reduction. Results obtained so far show that such reduced therapy does not result in more relapses, whereas side effects are drastically reduced.

The detection of is not only relevant in newly diagnosed patients with ALL, but also in ALL patients being treated after relapse. In addition, several studies have shown that measuring MRD in ALL patients who undergo a stem cell transplant is predictive for the clinical outcome (Goulden et al. 2003). Patients with relatively high levels of MRD before the transplant have a high risk of disease recurrence after the transplant (Fig. 6.5c). These patients may benefit from additional chemotherapy before the transplant, a more intense conditioning therapy, or immunotherapy shortly after the transplant (with the aim to induce a controlled "graft-versus-leukaemia" response).

6.4.6.2 Clinical Relevance of MRD Detection in Mature Lymphoid Malignancies

Measuring MRD is generally of little value in classically treated CLL patients, because these patients often still have a relatively high tumour load after treatment. However, in patients being treated with more recent (combinations of) drugs, measuring MRD is becoming relevant. Chemo-immunotherapy (for example, fludarabine, cyclophosphamide and rituximab, FCR) results in a subset of the CLL patients in MRD-negativity at the end of the treatment period and this seems to have predictive value for the clinical outcome in the long term. In recent clinical CLL studies evaluating new treatment modalities, the MRD result is currently being used as surrogate marker for clinical outcome.

In lymphoma patients, it is generally not possible to measure MRD in the tissue where the lymphoma developed (often a lymph node). In part of the lymphoma patients (especially FL and MCL) tumour cells can, however, also be detected in blood and/or bone marrow at the time of diagnosis. Molecular MRD studies may contribute to improved recognition of lymphoma patients with minimal infiltration of bone marrow (under the cytomorphological detection limit). In addition, MRD levels in bone marrow/blood in patients with leukemic NHL are related to the clinical outcome. Recent studies in FL show that the combination of chemotherapy and rituximab results in long-term complete remission in a subgroup of patients. These patients, in whom no MRD more could be detected in bone marrow and blood, had a significantly better survival than patients who were persistently MRD-positive or who became MRD-positive again. More than half of patients with MCL reached MRD-negativity (molecular remission) after intensive treatment. Patients with a molecular remission had longer response duration and achievement of a molecular remission was predictive of clinical outcome. Frequent analysis of MRD therefore provides a powerful predictor for the treatment outcome in patients with MCL.

6.4.7 *Molecular MRD Diagnostics with Other Markers*

In addition to IG/TCR gene rearrangements, other molecular markers may contribute to the diagnosis, classification and prognosis of haematological malignancies (Table 6.4). This is particularly relevant in chromosome abnormalities resulting in fusion genes, fusion gene transcripts, and/or abnormally expressed genes (Fig. 6.8).

Table 6.4 Chromosome abnormalities in lymphoid malignancies

Disease	Deviation	Target	Frequency (children/adult)	Prognostic impact
Precursor-B-ALL	t (9; 22)	*BCR-ABL*	5–8/30–35%	Poor
	t (1; 19)	*E2A-PBX1*	5–8/3–4%	None
	t (4; 11)	*MLL-AF4*	3–5[c]/3–4%	Poor
	11q23	*MLL deviation*	5–6/<5%	Poor
	t (12; 21)	*TEL-AML1*	~30/1–3%	Favourable
T-ALL	Del (1 p)	*SIL-TAL1*	10–25/5–10%	None
	t (8; 14)	*c-MYC-TCRA/D*	2–4%	?
	t (10; 14)	*HOX11-TCRD*	1–3%	?
FL	t (14; 18)	*IGH-BCL2*	80%	None
	3q27[a]	*BCL6*	5–10%	None
DLBCL	t (14; 18)	*IGH-BCL2*	20–30%	Poor
	3q27[a]	*BCL6*	30–40%	Favourable
MCL	t (11; 14)	*IGH-BCL1*	70%	None
Burkitt NHL	t (8; 14)	*MYC-IGH*	75–90%	Poor
	t (2; 8)	*IGK-MYC*		Poor
	t (8; 21)	*IGL-MYC*		Poor
Lymphoplasmacytic NHL	t (9; 14)[b]	*PAX5-IGH*	?	Poor
MALT-NHL	t (11; 18)	*API2/MALT1*	20%	Poor
ALCL	t (2; 5)	*NPM-ALK*	70–80%	Poor
	t (1; 2)	*TPM3-ALK*	10–20%	Poor

Symbols ? = unknown
[a]In at least 50% of the 3q27 aberrations (BCL6 locus) breaks in IG genes is involved (e.g. t(3;14); t (2;3), and t(3;22))
[b]Based on FISH experiments the frequency of t (9; 14) is at least 50% in lymphoplasmacytic NHL
[c]In infants with ALL the frequency of t (4; 11) is around 50%

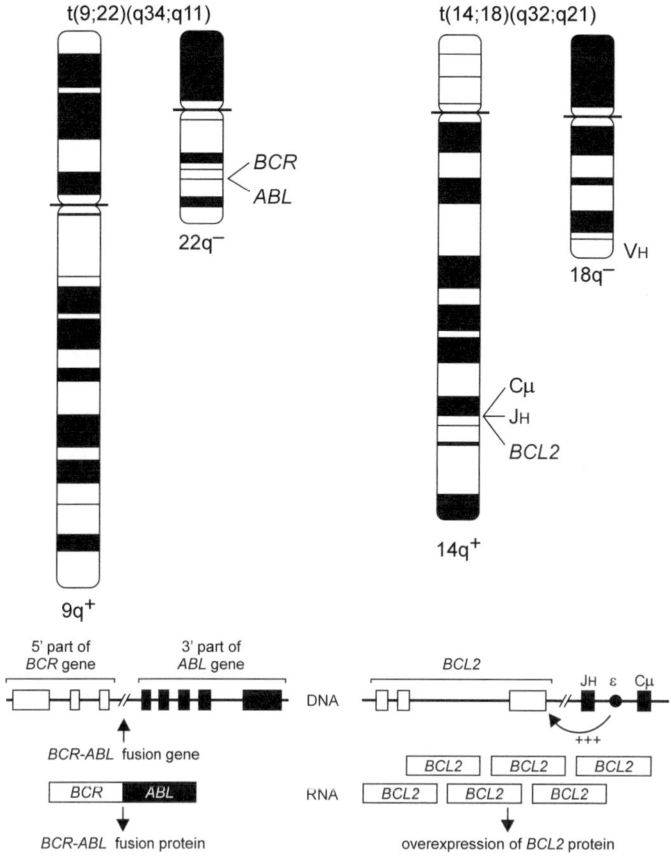

Fig. 6.8 Schematic representation of translocations leading to fusion genes, fusion transcripts and/or overexpression of genes. Chromosomal translocations can result in disturbed genes. If the break points are located in the coding sequences of the two genes, a fusion gene is formed consisting of a part of gene A and a part of gene B. Upon transcription and translation, fusion gene transcripts and fusion proteins are produced (*left panel*). Such a fusion protein may have contributed to the leukaemogenesis. If the break points are located in the regulatory sequences, transcription of gene A is controlled by the regulatory sequences of gene B; this will lead to an abnormal expression of gene A transcripts and proteins (*right panel*). Both types of translocations result in leukaemia-specific or patient-specific sequences that can be used as MRD-PCR target

These chromosome abnormalities can also be used to detect MRD during and after treatment. In approximately 40% of patients with ALL, fusion gene transcripts (RNA) can be detected. Standardized primer sets have been designed and are available for the detection of these fusion gene transcripts. In addition to their prognostic significance, fusion gene transcripts can also be used as MRD-PCR targets. Primer-probe sets for detecting the most common fusion gene transcripts are developed in the so-called "Europe Against Cancer" program. Depending on the type of fusion gene transcript, sensitivities between 10^{-3} and 10^{-6} can be obtained.

Such high sensitivities in combination with MRD-PCR targets that are not patient-specific may increase the risk of contamination, which is a significant pitfall.

In certain chronic leukaemias and lymphomas, chromosome aberrations can be detected. This includes the translocation t (14; 18) (BCL2 and IGH gene fusion) in FL, the translocation t(11;14) (BCL1 and IGH gene fusion) in MCL and translocations t (8;14), t (2;8), and t (8;22) (MYC and IG fusion) in Burkitt lymphomas. In the abovementioned translocations, the breaking points are generally outside the protein-coding regions and one of the partner genes comes under the influence of the IG promoter, resulting in an increased expression of the protein. This implies that for MRD analysis, these translocations can only be used at the DNA level (and not on the RNA level).

6.5 New Developments

Developments in molecular diagnostics are fast. In due time, there will be a number of new molecular techniques that can be applied in a research context in lymphoid malignancies, and that will find their way in the molecular diagnostics of lymphoid malignancies. This mainly relates to next generation sequencing (NGS) and digital droplet PCR (ddPCR) at the moment, which will be discussed in more detail below.

6.5.1 Next Generation Sequencing

Next generation sequencing (NGS) is a novel methodology in which large numbers of sequences can be generated (high-throughput sequencing) at the same time. These sequences concern genomic areas that are obtained through hybridization with probes (capture or pull-down strategy), or that are amplified in (multiplex) PCR strategies. Direct parallel sequencing of the areas of interest allows qualitative and quantitative analysis. In case of lymphoid malignancies, a lot of information can be obtained in this way about the exact IG/TCR gene rearrangements in leukaemias and/or lymphomas, including a detailed view of the IG/TCR repertoire of normal cells. In addition, the NGS method is reasonably fast, whereas it is expected that costs will further drop in the coming years. However, good bioinformatics support is essential for accurate and reliable analysis of the data. Despite the fact that the IG/TCR NGS technology is rapidly developing and that proof-of-principle studies for clonality testing and MRD analysis have been published, there are still a number of issues that need extra attention before the IG/TCR NGS technology is routinely applicable:

- With such highly sensitive analyses, the exact meaning of identical IG/TCR rearrangements for the interpretation of (mono) clonality will have to be redefined

- For a quantitative analysis of the repertoire of IG/TCR sequences, it is crucial that the different primers show a highly comparable efficiency. Especially in lymphoid cells this is challenging, because:
  - due to the wide variety of V, D and J genes, many (family) primers are used that do not show 100% complementary with all family members
  - somatic hyper mutation will influence primer attachment to the IG genes
- The IG/TCR gene rearrangements will, in principle, only be present in lymphoid cells, while the DNA (or cDNA) is derived from all cells. This may make it difficult to quantify the sequences in a proper way. By spiking the DNA with different amounts of known sequences, a post-run correction can be performed
- It is claimed that the IG/TCR NGS methodology achieves sensitivities of 10^{-6}, but in many cases this is probably overestimated. Such sensitivities can only be achieved if DNA of at least 1 million cells is being analysed (corresponding to about 3 μg), which in practice is often not feasible.

6.5.2 *Digital Droplet PCR*

In digital droplet PCR (ddPCR), the PCR reaction is more or less divided into thousands of mini-PCR reactions (in droplets). This is done in such a way that every drop contains approximately one piece of template DNA (or cDNA). After the PCR reaction, the DNA content is measured per drop, allowing for a qualitative and quantitative analysis. This method is particularly interesting for IG/TCR—based MRD analyses, and currently available data indicate that ddPCR is indeed very robust and reproducible. The big advantage of the ddPCR method is that quantitative data can be obtained, even if no baseline material is available to generate a standard curve. Although in patients with ALL this will rarely be the case, it is a great advantage in lymphoma patients (with often more limited amounts of baseline material). Also, the ddPCR will have to be further optimized and validated, before robust clinical application is possible in lymphoid malignancies.

6.6 Conclusions

Molecular diagnostics has obtained a prominent position in the whole diagnostic spectrum of haematological malignancies. Rearranged IG/TCR genes are DNA markers which are present in virtually all lymphoid malignancies and can be used for clonality testing purposes at initial diagnosis. In addition, IG/TCR gene rearrangements are used as MRD-PCR targets for monitoring therapy response. Another application concerns IGHV SHM analysis in CLL patients, in which the SHM status is important as prognostic marker and can be used for inclusion and potentially also for stratification of patients.

In addition to IG/TCR gene rearrangements, several other targets can be used for molecular diagnostics in patients with haematological malignancies. This concerns fusion genes, fusion gene transcripts, and genes that are aberrantly expressed. The wide availability of molecular targets and the easy availability of blood (and to a lesser extent bone marrow) make the haematological malignancies a paradigm for technical developments that can also be implemented in other types of cancer. This also applies to the latest molecular techniques such as NGS and ddPCR that have already proven their value for lymphoid malignancies in a research context.

References

Baliakas P, Hadzidimitriou A, Sutton LA, Minga E, Agathangelidis A, Tsanousa A, Scarfo L, Davis Z, Yan XJ, Shanafelt T, Plevova K, Sandberg Y, Vojdeman FJ, Boudjogra M, Tzenou T, Chatzouli M, Chu CC, Veronese S, Gardiner A, Mansouri L, Smedby KE, Bredo Pedersen L, Moreno D, van Lom K, Giudicelli V, Shkurova Francova H, Nguyen-Khac F, Panagiotidis P, Juliusson G, Angelis L, Anagnostopoulos A, Lefranc MP, Trentin L, Catherwood M, Montillo M, Geisler G, Langerak AW, Pospisilova S, Chiorazzi N, Oscier D, Jelinek D, Darzentas N, Belessi C, Davi F, Rosenquist R, Ghia P, Stamatopoulos K. Clinical effect of stereotyped B-cell receptor immunoglobulins in chronic lymphocytic leukaemia: a retrospective multicentre study. Lancet Hematol. 2014;1:e74–84.

Damle RN, Wasil T, Fais F, Ghiotto F, Valetto A, Allen SL, Buchbinder A, Budman D, Dittmar K, Kolitz J, Lichtman SM, Schulman P, Vinciguerra VP, Rai KR, Ferrarini M. Chiorazzi N Ig V gene mutation status and CD38 expression as novel prognostic indicators in chronic lymphocytic leukemia. Blood. 1999;94:1840–7.

Flohr T, Schrauder A, Cazzaniga G, Panzer-Grümayer R, van der Velden V, Fischer S, Stanulla M, Basso G, Niggli FK, Schäfer BW, Sutton R, Koehler R, Zimmermann M, Valsecchi MG, Gadner H, Masera G, Schrappe M, van Dongen JJ, Biondi A, Bartram CR. International BFM Study Group (I-BFM-SG). Minimal residual disease-directed risk stratification using real-time quantitative PCR analysis of immunoglobulin and T-cell receptor gene rearrangements in the international multicenter trial AIEOP-BFM ALL 2000 for childhood acute lymphoblastic leukemia. Leukemia. 2008;22:771–82.

Ghia P, Stamatopoulos K, Belessi C, Moreno C, Stilgenbauer S, Stevenson F, Davi F, Rosenquist R. ERIC recommendations on IGHV gene mutational status analysis in chronic lymphocytic leukemia. Leukemia. 2007;21:1–3.

Goulden N, Bader P, van der Velden V, Moppett J, Schilham M, Masden HO, Krejci O, Kreyenberg H, Lankester A, Révész T, Klingebiel T, van Dongen J. Minimal residual disease prior to stem cell transplant for childhood acute lymphoblastic leukaemia. Br J Haematol. 2003;122:24–9.

Hamblin TJ, Davis Z, Gardiner A, Oscier DG. Stevenson FK Unmutated Ig V(H) genes are associated with a more aggressive form of chronic lymphocytic leukemia. Blood. 1999;94:1848–54.

Langerak AW, Groenen PJTA, van Krieken JHJM, van Dongen JJM. Ig/TCR clonality diagnostics. Exp Opin Med Diagn. 2007;1:451–61.

Langerak AW, Davi F, Ghia P, Hadzidimitriou A, Murray F, Potter KN, Rosenquist R, Stamatopoulos K, Belessi C. Immunoglobulin sequence analysis and prognostication in CLL: guidelines from the ERIC Review Board for reliable interpretation of problematic cases. Leukemia. 2011;25:979–84.

Langerak AW, Groenen PJTA, Brüggemann M, Beldjord K, Bellan C, Bonello L, Boone E, Carter I, Catherwood M, Davi F, Delfau-Larue MH, Diss T, Evans PAS, Gameiro P, Garcia

Sanz R, Gonzalez D, Grand D, Håkansson A, Hummel M, Liu H, Lombardia L, Macintyre EA, Milner B, Montes-Moreno S, Schuuring E, Spaargaren M, Hodges E, van Dongen JJM. EuroClonality/BIOMED-2 guidelines for interpretation and reporting of Ig/TCR clonality testing in suspected lymphoproliferations. Leukemia. 2012;26:2159–72.

Pongers-Willemse MJ, Seriu T, Stolz F, d'Aniello E, Gameiro P, Pisa P, Gonzalez M, Bartram CR, Panzer-Grümayer ER, Biondi A, San Miguel JF, van Dongen JJ. Primers and protocols for standardized detection of minimal residual disease in acute lymphoblastic leukemia using immunoglobulin and T cell receptor gene rearrangements and TAL1 deletions as PCR targets: report of the BIOMED-1 CONCERTED ACTION: investigation of minimal residual disease in acute leukemia. Leukemia. 1999;13:110–8.

van Dongen JJ, Seriu T, Panzer-Grümayer ER, Biondi A, Pongers-Willemse MJ, Corral L, Stolz F, Schrappe M, Masera G, Kamps WA, Gadner H, van Wering ER, Ludwig WD, Basso G, de Bruijn MA, Cazzaniga G, Hettinger K, van der Does-van den Berg A, Hop WC, Riehm H, Bartram CR. Prognostic value of minimal residual disease in acute lymphoblastic leukaemia in childhood. Lancet. 1998 Nov 28;352(9142):1731–8.

van Dongen JJM, Langerak AW, Brüggemann M, Evans PA, Hummel M, Lavender L, Delabesse E, Davi F, Schuuring E, Garcia Sanz R, van Krieken JHJM, Droese J, Gonzalez D, Bastard C, Spaargaren M, San Miguel JF, Parreira A, Smith JF, Morgan G, Kneba M, Macintyre EA. Design and standardization of PCR primers and protocols for detection of clonal immunoglobulin and T-cell receptor gene rearrangements in suspect lymphoproliferations. Report of the BIOMED-2 Concerted Action BMH4-CT98-3936. Leukemia. 2003;17:2257–317.

van Krieken JHJM, Langerak AW, Macintyre EA, Kneba M, Smith JL, Garcia Sanz R, Morgan GJ, Parreira A, Molina T, Cabeçadas J, Gaulard P, Jasani B, Garcia JF, Ott M, Hannsmann ML, Berger F, Hummel M, Davi F, Brüggemann M, Lavender FL, Schuuring EMD, Evans PAS, White H, Salles G, Groenen PJTA, Gameiro P, Pott Ch, van Dongen JJM. Improved reliability of lymphoma diagnostics via PCR-based clonality testing. Report of the BIOMED-2 Concerted Action BHM4-CT98-3936. Leukemia. 2007;21:201–6.

van der Velden VH, Hochhaus A, Cazzaniga G, Szczepanski T, Gabert J, van Dongen JJ. Detection of minimal residual disease in hematologic malignancies by real-time quantitative PCR: principles, approaches, and laboratory aspects. Leukemia. 2003;17:1013–34.

van der Velden VH, Panzer-Grümayer ER, Cazzaniga G, Flohr T, Sutton R, Schrauder A, Basso G, Schrappe M, Wijkhuijs JM, Konrad M, Bartram CR, Masera G, Biondi A, van Dongen JJ. Optimization of PCR-based minimal residual disease diagnostics for childhood acute lymphoblastic leukemia in a multi-center setting. Leukemia. 2007;21:706–13.

van der Velden VH, Cazzaniga G, Schrauder A, Hancock J, Bader P, Panzer-Grumayer ER, Flohr T, Sutton R, Cave H, Madsen HO, Cayuela JM, Trka J, Eckert C, Foroni L, Zur Stadt U, Beldjord K, Raff T, van der Schoot CE, van Dongen JJ. European Study Group on MRD detection in ALL (ESG-MRD-ALL). Analysis of minimal residual disease by Ig/TCR gene rearrangements: guidelines for interpretation of real-time quantitative PCR data. Leukemia. 2007b;21:604–11.

Chapter 7
Molecular Diagnostic Testing in Clinical Chemistry

Daan van den Broek, Eric Vermeer, Dorine Swinkels and Ron H.N. van Schaik

Abstract Initially, DNA analysis was restricted to the identification of inborn errors of disease and was performed in clinical genetic centres. Nowadays, DNA analysis is increasingly used also in clinical chemistry as biomarkers for detecting haemostatic parameters, hereditary hemochromatosis, hemoglobinopathies, and the classification and treatment of hematologic malignancies. Specifically the use of DNA analysis to explain, and to predict response to medication (Pharmacogenetics) is taking an increasingly important role. In addition, the identification of cell free DNA, detecting drug therapy resistance mutations (like EGFR p.T790M for lung cancer) is a new and rapidly evolving field to drug treatment. New horizons will expectedly soon open with the availability of advanced sequencing technology, pertaining to the analysis of small (amplified) fragments to the whole genome (WGS).

Keywords Pharmacogenetics · Hematologic malignancies · Haemoglobinopathy · Hereditary hemochromatosis · Haemostatic parameters

7.1 Introduction

Laboratory diagnostic testing on DNA has evolved rapidly over the last 15 years. Initially, DNA analysis was restricted to the identification of inborn errors of disease. Well-known examples are DNA analysis for cystic fibrosis (CF) and

D. van den Broek
NKI, Amsterdam, The Netherlands

E. Vermeer
Results Laboratory, Dordrecht, The Netherlands

D. Swinkels
Radboud UMC, Nijmegen, The Netherlands

R.H.N. van Schaik (✉)
Erasmus MC, Rotterdam, The Netherlands
e-mail: r.vanschaik@erasmusmc.nl

E. van Pelt-Verkuil et al. (eds.), *Molecular Diagnostics*,
DOI 10.1007/978-981-10-4511-0_7

Duchenne muscular dystrophy. Nowadays, laboratory diagnostic testing plays an important role in identifying patients at risk, assisting in the identification of individuals that harbour DNA variants that have been associated with an increased risk of disease. This can be blood clotting -related genes, such as Factor V Leiden and Factor II, giving an increased risk on clotting, or lipid metabolism-related gene variants, such as ApoE, affecting the risk on cardiovascular events. DNA variations in genes associated with iron overload diseases (hemochromatosis) may predispose patients to organ failure, and pharmacogenetic analyses will identify a patient risk to side effects on drug therapy. Very recently, the use of cell-free DNA as a new biomarker for the detection of tumour resistance or for transplant rejection has experienced an increase in popularity.

Whereas DNA analysis initially depended very much on laboratory-developed tests (LDTs), currently, CE-marked and/or FDA-marked molecular tests are available and are implemented in the clinical laboratory. For many DNA tests, external quality control programs are operational (SKML (The Netherlands), RfB (Germany), EQAS (UK)).

To give an impression of the current applications in the Clinical Chemistry field, the topics listed below will be highlighted:

- Section 7.2 Molecular diagnosis of haemostatic parameters
- Section 7.3 Molecular diagnosis of hereditary hemochromatosis
- Section 7.4 Molecular diagnostics for pharmacogenetics
- Section 7.5 Haemoglobinopathy
- Section 7.6 Molecular diagnostics in hematologic malignancies
- Section 7.7 Molecular diagnosis of rare genetic abnormalities
- Section 7.8 Molecular diagnostics by means of cell-free DNA
- Section 7.9 Future developments.

7.2 Molecular Diagnostics of Haemostatic Parameters

Haemostasis is an essential process in the vascular system. It serves to stop bleeding at the site of an injury and to prevent blood loss by the formation of a haemostatic plug, including maintenance of this plug once the wound has healed. A deficiency or enhancement of the activity of one of the components of this system can result in a bleeding tendency or thrombosis, respectively. At least five different components are involved in the haemostatic process: blood vessels, platelets, plasma coagulation factors and their inhibitors and the fibrinolytic system. The clinical chemistry laboratory plays an important role in assessing platelet function through the measurement of aggregation characteristics, the speed of clot formation (prothrombin time and activated partial thromboplastin time), and the detection of coagulation factor deficiencies (e.g., factor VIII in haemophilia A). An important application lies in the identification of abnormalities at the DNA level that are strongly correlated with improper functioning of the coagulation cascade.

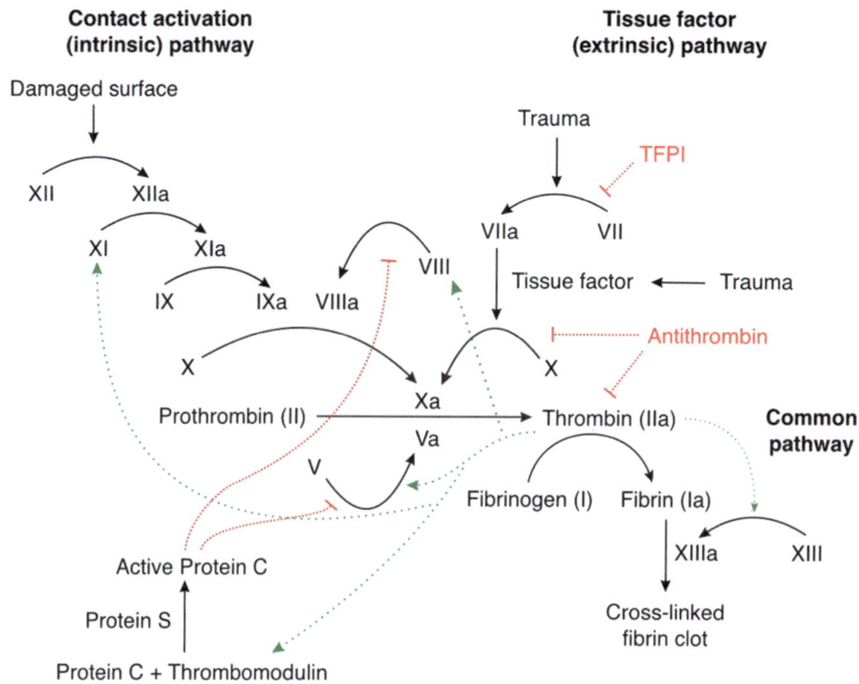

Fig. 7.1 Coagulation cascade, in which the activated clotting factors finally result in a cross-linked fibrin polymer network

Activated protein C (APC) is a serine proteinase with strong anticoagulant properties. APC is formed from an inactive precursor. In normal haemostasis, APC is responsible for the proteolytic inactivation of Factors Va and VIIIa, in the presence of protein S as a cofactor, thus inhibiting the activation of coagulation (Koeleman 1994) (Fig. 7.1).

Resistance to digestion by APC is associated with a single nucleotide mutation in the gene coding for Factor V. The responsible mutation is the replacement of an adenine base (A) by a guanine base (G) (G1691A) at position 1691 of the factor V gene. This results in the disappearance of a restriction site for Mnl-1 in a 220 bp Factor V PCR product and in the appearance of a 153 base pair fragment after digestion with Mnl-1 (Fig. 7.2). At the protein level, this mutation encodes an amino acid substitution at position 506 of Factor V: arginine (R) is replaced by glutamine (Q) (R506Q) (Bertina et al. 1994). The Factor Va mutation is located at the site of one of the inactivating cleavage sites for APC. Due to this mutation, APC can no longer cleave Factor Va, making Factor Va APC resistance and thus increasing coagulation. Patients with this Factor V mutation (Factor V-Leiden), have an increased risk of venous thrombosis (thrombophilia). Individuals who are experiencing venous thrombosis show a high prevalence of this mutation (20–50%), when other causes such as immobilization, recent surgery, pregnancy, the use of oral contraceptives or malignancy are excluded (Aillaud et al. 1995).

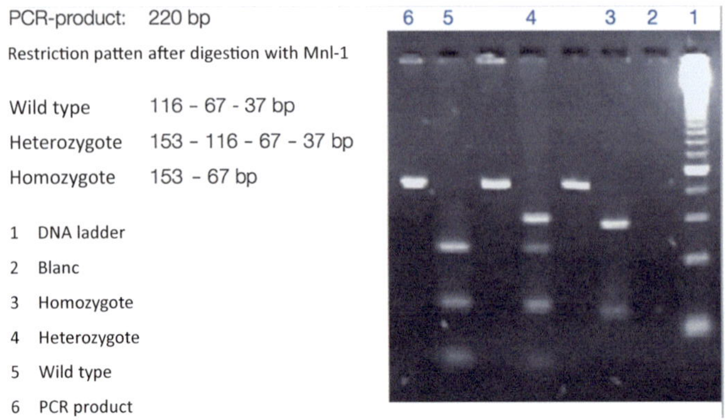

Fig. 7.2 Agarose gel with DNA fragments after restriction enzyme treatment with Mnl-1 of the amplified Factor V DNA fragment. Indicated are the patterns of a homozygote (*3*), heterozygote (*4*), and a wild type (*5*) patient. NB: the 37 bp fragment is not visible on this agarose gel

Approximately 90% of cases of APC resistance are due to the Factor V Leiden mutation (R506Q). Furthermore, women with the Factor V Leiden mutation have an increased risk of pre-eclampsia during pregnancy (Grandone et al. 1997).

APC-resistance is demonstrated by means of a plasma-APC-resistance test, measuring clotting times (APTT) in the presence and absence of exogenous APC. DNA analysis for a point mutation in Factor V is subsequently performed in most laboratories whenever APC-resistance (normalized APC sensitivity ratio <0.84) is detected. The point mutation in the DNA demonstrating the Factor V Leiden (R506Q) mutation indicates a genetically determined APC-resistance. Comparable, non-DNA related clinical coagulation problems can also be acquired during lifetime. For example, patients who are pregnant, are taking oral contraceptives or have been diagnosed with lupus anticoagulant (LAC) can also develop APC resistance or reduced susceptibility to APC. It is obvious that the acquired form of APC-resistance cannot be demonstrated using the Factor V-Leiden molecular test. A careful evaluation of the patient and the interpretation of the laboratory results are therefore essential for correct diagnosis and treatment.

7.3 Molecular Diagnosis of Hereditary Hemochromatosis

The balance of iron in the body is determined by several factors (Fig. 7.3). Hereditary hemochromatosis (HFE) is an autosomal recessive hereditary disease characterized by iron accumulation in tissues and organs. The diagnosis is usually made around the age of fifty and is based on loss of function of one or more organs. The clinical picture can vary greatly. Most common are accumulation in liver (cirrhosis), pancreas (diabetes), heart muscle (cardiomyopathy), testicles (infertility), and joints (arthritis). A diagnosis at the pre-symptomatic stage is very

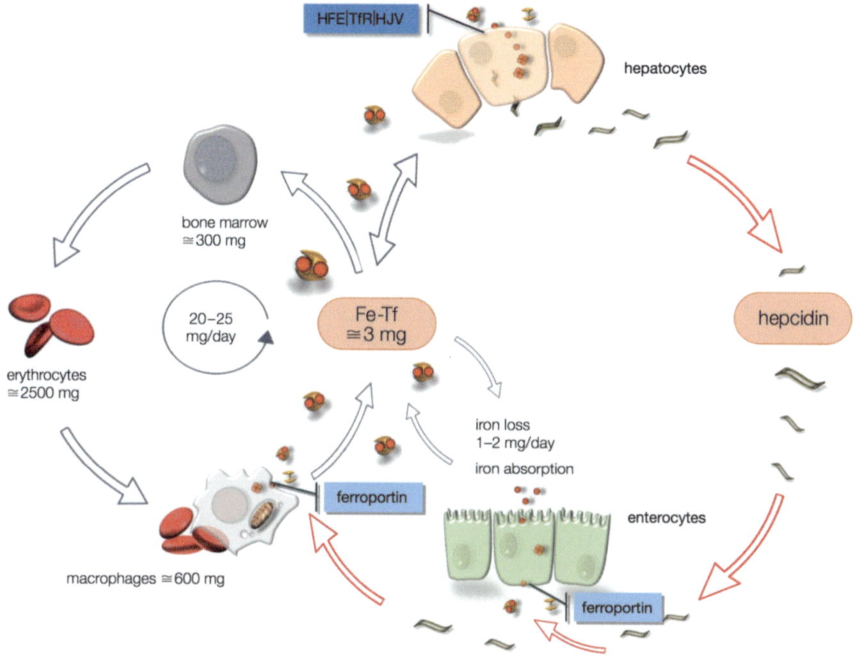

Fig. 7.3 Iron cycle. Hepcidin, produced in hepatocytes, is key in the resorption of iron. This hormone regulates the iron resorbing proteins in the basolateral cells of the intestine and in the macrophages. Low hepcidin levels results in a high iron resorption

important. Irreversible damage caused by iron overload can be prevented relatively easily by periodical phlebotomies (Dutch guidelines 2007; EASL 2010; Porto 2016; van Bokhoven 2011). Parameters like iron saturation and ferritin concentration (in a non-infectious situation, so in the absence of elevated CRP) are good indicators of iron overload, but cannot be used to distinguish between the primary and secondary forms, such as those made by stacking anaemia's and in patients receiving multiple transfusions.

In 1996, it was demonstrated that the c.845G>A point mutation (rs 1800562) in the HFE gene is associated with hereditary hemochromatosis (Feder et al. 1996). The protein encoded by the HFE gene, which is located on chromosome 6, forms a dimer with β2-microglobulin. This transmembrane protein complex is present in the membranes of intestinal mucosa cells, such as macrophages, and hepatocytes. The function of the HFE protein is not well understood. It is thought to be involved in the signalling of high circulating iron levels by the hepatocytes and the subsequent inhibition of iron resorption from the intestine. The c.845G>A mutation in the HFE protein results in a cysteine to tyrosine substitution (p.Cys282Tyr). Due to this mutation, a sulphur bridge, which is essential for the three-dimensional structure of the protein and the formation of the dimer, cannot be formed anymore (Fig. 7.4).

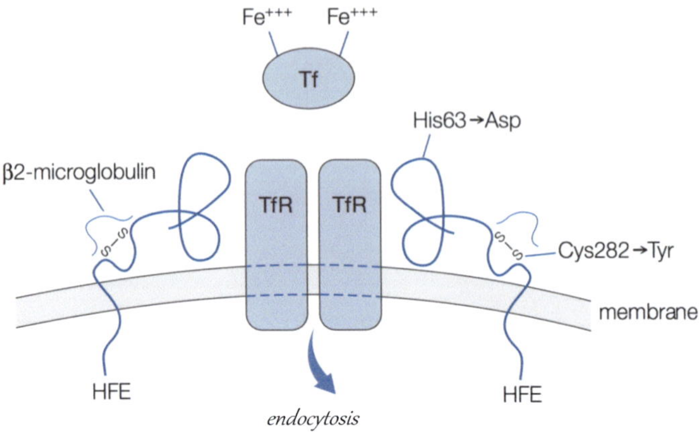

Fig. 7.4 Schematic representation of the HFE protein. HFE forms a dimer with b2-microglobulin that can facilitate the uptake of the transferrin-bound iron

The importance of molecular diagnostics in this field can best be explained with a case example: a 52-year-old man that had been suffering from fatigue and vague malaise for several years was admitted with progressive liver disorders. From routine clinical chemistry analysis, both a high ferritin concentration and a high iron saturation of transferrin were found in the blood. Based on the suspicion of iron overload disease, DNA testing was requested, which then confirmed the diagnosis (Dutch guidelines 2007; EASL 2010; Porto 2016; van Bokhoven 2011).

A specific HFE primer pair, amplifying a 390 bp fragment, normally comprises one constitutional Rsa I restriction site. Fragments of 140 bp and 250 bp are formed after treatment of the PCR fragment with this enzyme. The c.854G>A mutation creates a second restriction site, generating an additional DNA fragment in the 390 bp PCR product on gel electrophoresis.

Sanger sequencing of exons 2 and 4 of the HFE gene in a 48-year-old sister of the patient discussed above demonstrated the presence of the c.854G>A mutation in homozygous form. Subsequent additional clinical chemistry analyses showed increased iron saturation in the sister and an increased serum ferritin concentration. Yet, no clinical symptoms of iron overload were present. This defect was thus identified preclinically, using DNA analysis in a family member of an affected individual. By performing regular phlebotomies, loss of organ function as a clinical consequence of iron overload could be prevented.

There are also indications that the p.His63Asp mutation (c.187c>G; rs1799945) in combination with the p.Cys282Tyr mutation on the other chromosome (compound heterozygote) may cause a mild form of iron overload. The increase in serum iron parameters in these patients is associated with concomitant disorders, such as obesity, hepatic steatosis, and alcohol abuse and in itself rarely leads to tissue damage (Porto 2016). In 7% of the patients with hemochromatosis, this compound heterozygosity for C282Y/H63D was involved (Feder et al. 1997).

7.4 Molecular Diagnostics for Pharmacogenetics

Pharmacogenetics is the field that investigates changes in the DNA that are related to drug metabolizing enzymes and drug transporters, which may explain the different responses to drug therapies that are seen in patients. Moderne Vogel introduced the term Pharmacogenetics in 1959. We would like to highlight three clear historical examples of the link between DNA variations and the susceptibility to drugs:

1. In the sixties, it was apparent that a genetic component played a role in the ability to acetylate the drug isoniazid, separating the population in slow and rapid acetylators. This was caused by genetic polymorphisms in the gene coding for N-acetyltransferase 2 (NAT-2) (Evans et al. 1960, 1969).
2. Patients suffering from pseudocholinesterase deficiency have an extremely long reaction to the muscle relaxants succinylcholine and mivacurium, with muscle relaxation for up to 6 h, instead of 5–10 min in individuals with normal levels of the protein. This phenomenon, occurring in 1:1250 patients, is caused by the incapability of these patients to degrade succinylcholine and mivacurium. In 50% of these cases, a DNA variant in the butyrylcholinesterase (BChE) gene, encoding for the enzyme pseudocholinesterase is responsible.
3. A third well-known example of pharmacogenetics deals with the enzyme thiopurine-S-methyl transferase (TMPT). This enzyme metabolizes 6-mercaptopurine, used in treating Acute Lymphatic Leukaemias (ALL), and azathioprine, used for the treatment of psoriasis, Crohn's disease, rheumatoid arthritis and solid organ transplantation (Fig. 7.5). When analysing the enzyme

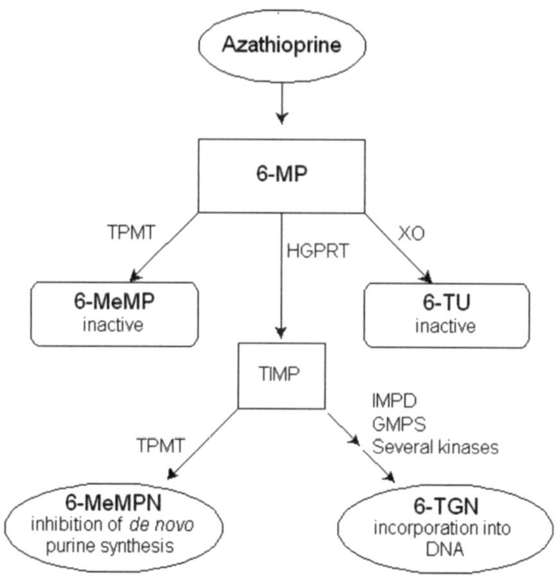

Fig. 7.5 Metabolism of azathioprine and 6-mercaptopurin (6-MP). Azathioprine is converted to 6-MP. 6-MP is partially metabolised to the active thioguanine nucleotides (6-TGNs) and to the inactive 6-MeMP. This latter reaction is catalyzed by TPMT. When TPMT activity is absent, due to for instance a genetic polymorphisms, the full dose of Aza or 6-MP will be concerted to the 6-TGNs, giving potentially lethal bone marrow suppression

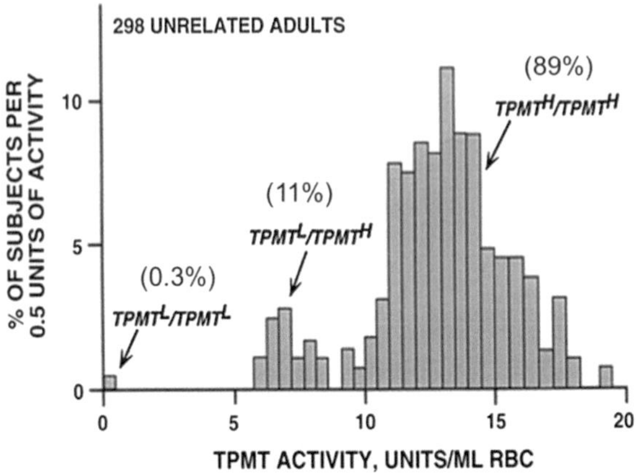

Fig. 7.6 Distribution of TPMT enzyme activity in the population. Shown is a trimodal distribution, in which 89% has a high activity (two active TMPT alleles), 11% has an intermediate metabolism (1 active, 1 inactive TPMT allele) and 0.3% is deficient in TPMT activity (2 inactive TPMT alleles)

activity in the population, a trimodal distribution became apparent, pointing to a genetic component (Fig. 7.6) (Weinshilboum and Sladek 1980). Indeed, three single nucleotide polymorphisms (SNPs) could be identified that affected enzyme activity: G238C, G460A and A719C (Fig. 7.7). These SNPs, corresponding to the low activity variant alleles *2 (G238C), *3A (G460A, A719C) and *3C (A719C), were found to be able to explain the enzymatic activity of TPMT in 95% of the cases (Schaeffeler et al. 2004). When identified, heterozygote individuals can be treated with 50% of the standard dose whereas homozygotes should receive only 10% of the standard dose. When TPMT deficient patients are treated with normal dosages, lethal consequences have been described. Indeed, the information on TPMT deficiency has been taken up in the drug label.

The examples above describe specific gene-drug combinations. Yet, more interesting perhaps, are the enzymes of the Cytochrome P450 (CYP450) family. These enzymes are involved in the metabolism of eighty percent of all drugs. The system consists of several sub-enzymes, such as CYP1A2, CYP2B6, CYP2C9, CYP2C19, CYP2D6, CYP3A4, and CYP3A5, each having their own substrate specificity (Fig. 7.8). The CYP3A4 enzyme is the most abundant enzyme and is involved in the metabolism of 30–40% of all drugs. CYP2D6, which at the protein level consists of only 2% of the total CYP450 protein, is involved in the metabolism of 20–25% of commonly prescribed drugs, such as antidepressants, antipsychotics, beta blockers and opioids.

The activity of these enzymes is not the same among individuals. For CYP2D6, 5–10% of the population is classed as a slow or poor metabolizer (PM), 15% is classed as an intermediate (IM) and 3% as an ultra rapid metabolizer

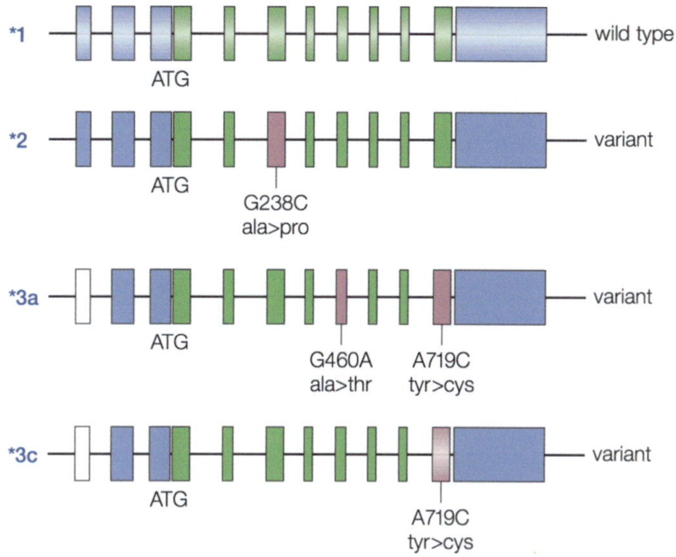

Fig. 7.7 TPMT wild type (*1) and three main variant alleles, corresponding with low enzymatic activity. The *2 allele is characterized by the single nucleotide polymorphism (SNP) 238G>C, the *3A allele by the presence of the two SNPs 460G>A and 719A>C, (main variant allele in caucasian population) whereas the *3C allele (main variant allele in Asain populations) is characterized by only the 719A>C SNP

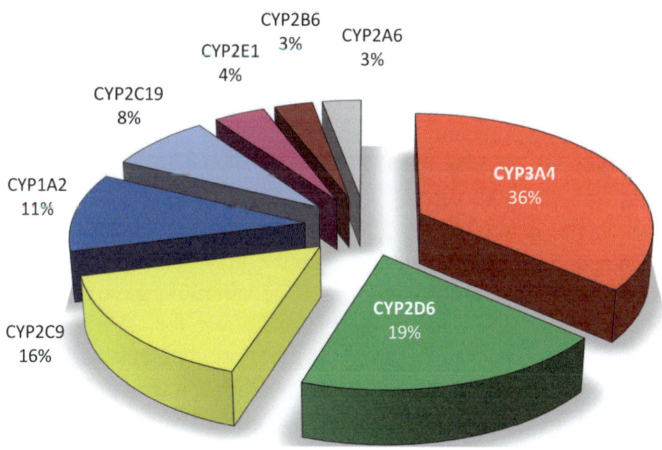

Fig. 7.8 The CYP450 enzyme system, showing the different sub-enzymes. In brackets the percentage of drugs being metabolized by each enzyme

(UM) (Fig. 7.9). These percentages may differ between ethnicities. Also here, genetics proved to be an important factor to explain this distribution. Patients with two inactive alleles are poor metabolizers (PM), whereas patients with one active

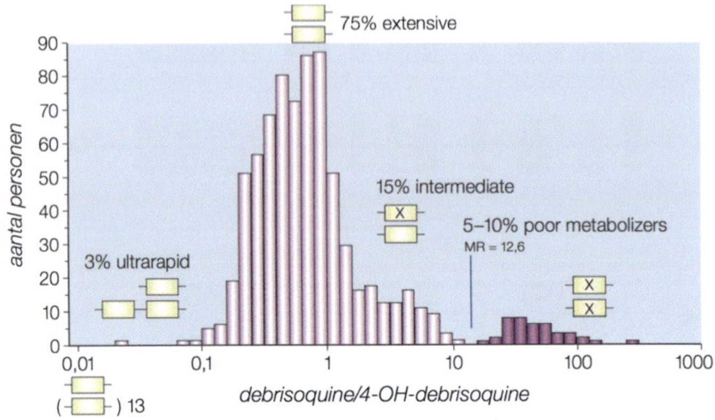

Fig. 7.9 Distribution of the CYP2D6 enzymatic activity in the Caucasian population, showing a poor, intermediate, extensive, (normal) and ultrarapid metabolizer group

and one inactive allele are intermediate metabolizers (IM). Individuals without DNA variants, thus expected to have normal enzyme activity, are called "extensive", or normal metabolizers (EM). Ultra-rapid metabolizers (UEM) for CYP2D6 also exist, due to the presence of gene duplication, giving these patients excess CYP2D6 enzyme. Genotyping for CYP2D6 can thus give information whether drug dosing for 20–25% of drugs should be adjusted. For CYP2C19, ultra-rapid metabolism is caused by another phenomenon, being the inheritance of two hyperactive *CYP2C19*17* alleles (Sim et al. 2006). The clinical consequences of a predicted PM phenotype are that the patient experiences higher concentrations of the parent compound and thus potentially adverse drug reactions when given a standard dose. The UEM-phenotype, in contrast, will demonstrate a lack of therapeutic effect. When the enzymes are involved in drug activation (such as codeine, tramadol, and tamoxifen), it is the CYP2D6 PM that will have a phenotype that does not show any effect on therapy.

The nomenclature of the variant alleles uses a star "*" classification. The most common DNA sequence which encodes an active enzyme is known as the *1 allele. Subsequently, chronological discovery of genetic variations gave rise to: *2, *3, *4, etc. alleles. Nowadays, 105 variant alleles are known for CYP2D6 (www.cypalleles.ki.se). The *CYP2D6*4* allelic variant is the most abundant variant allele in the Caucasian population (a frequency 18%), followed by the total deletion of the gene: *CYP2D6*5* (11%). For CYP2D6, there are also alleles with reduced activity, such as *41. This variant still generates some enzymatic activity, but less than the wild type. It is important to realize that the genotype *1/*1 is a default value. If no variations are found, the genotype is said to be *1/*1. The predictive value thus depends on the number of variants examined. For CYP2D6, for instance, our laboratory (Erasmus MC, Rotterdam, the Netherlands) examines 15 variants. Of course, this may differ between laboratories. There are CE-IVD approved tests available for CYP2D6, which enables the testing of multiple variant alleles at once, for instance using DNA chips (Fig. 7.10).

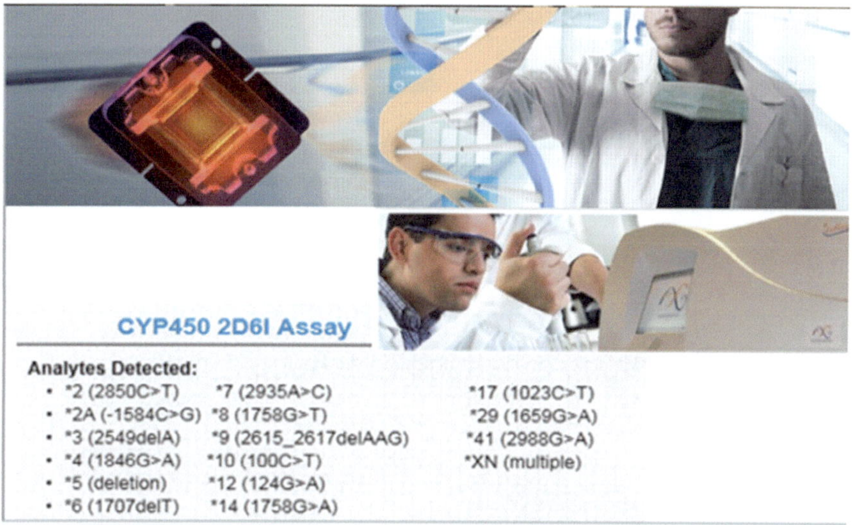

Fig. 7.10 Example of a CYP2D6 genotyping platform (CE-IVD), allowing the detection of 16 CYP2D6 genetic variants. For each allele, it is indicated which characterising SNP is analysed

One example of CYP450 testing in clinical practice is the use of a CYP2D6 assay for antidepressants, antipsychotic drugs and beta-blockers, since poor metabolizers may experience severe side effects when these drugs are given in normal dosages. On the other hand, tamoxifen and the pain medication drugs codeine and tramadol depend on CYP2D6 for activation of the drug. Here, CYP2D6 PM will not benefit fully from the anticancer effect of tamoxifen, or the pain killing effect of codeine. The CYP2C19, with 2–3% PMs, also plays a role in anti-depressant metabolism, but additionally has an important clinical role in the activation of the platelet aggregation inhibitor clopidogrel (Plavix). Carriers of a deficient allele (e.g. *CYP2C19*2*) will benefit less of clopidogrel and should therefore preferably be prescribed another drug, like prasugrel or ticagrelor. Information on this pharmacogenetic interaction has been included by the FDA in the package leaflet of clopidogrel as a "boxed warning" (Table 7.1).

An important aspect is how to make this genetic information clinically useful. In fact, the Dutch Working Group on Pharmacogenetics, initiated by the Royal Dutch Society of Pharmacy (KNMP) in 2005, started to rate all evidence concerning pharmacogenetics and the effects on plasma concentrations and clinical outcomes. This has resulted in DNA-based dosing guidelines for over 80 drugs (Swen et al. 2011). These guidelines are available to all pharmacists in The Netherlands, and can also be accessed internationally through www.pharmgkb.org.

Genotyping has advantages over measuring drug concentrations: it is prospective, it is a test that needs to be performed only once and can be performed on only a small amount of blood of the patient. The disadvantage of DNA testing is that it is only a prediction of the phenotype and thus of the drug exposure, and it does not

Table 7.1 Enzyme variants involved in drug metabolism

Enzyme	PM	IM	EM	UEM	Drugs
CYP2C9	2%	20%	88%	0%	Acenocoumarol, phenprocoumon, warfarin
CYP2C19	2%	20%	70%	8%	Clopidogrel (A), antidepressants
CYP2D6	8%	30%	60%	2%	Antidepressants, antipsychotics, codeine (A), tramadol (A), tamoxifen (A), metoprolol
TPMT	0.3%	10%	90%	0%	6-mercaptopurine, azathioprine
DPYD	2%	14%	84%	0%	Capecitabine, 5-FU
BChE	0.3%	3%	97%	0%	Succinylcholine, mivacurium
HLA-B*5701	4% NEG		96% POS		Abacavir
VKORC1	20% sensitive		80% normal		Acenocoumarol, phenprocoumon, warfarin

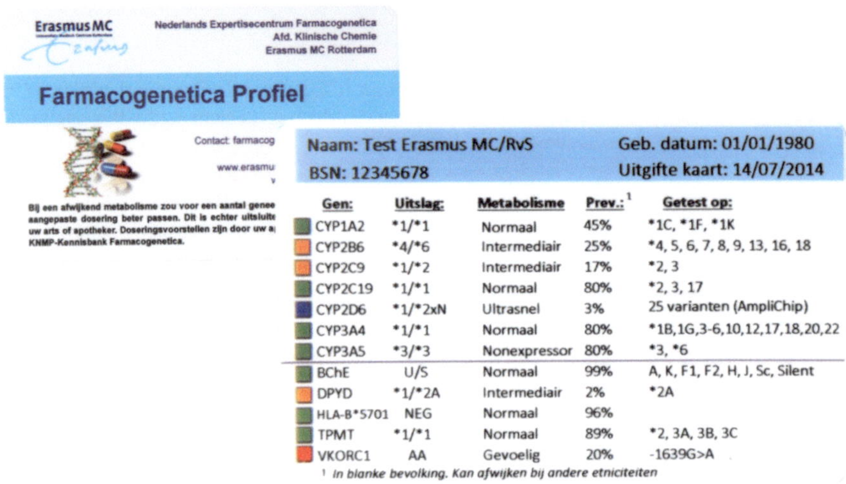

Fig. 7.11 DNA passport for medication, as being given out to patients since January 1, 2016, by the International Reference centre Pharmacogenetics, Dept. Clinical Chemistry, Erasmus MC Rotterdam, The Netherlands. In the Netherlands, patients can present this card to any pharmacist, who has access to genotype dependent dosing advices for over 80 drugs, provided by the Dutch Working Group PGx (DWGP) from the Royal Dutch Pharmacist Association (KNMP)

actually measure it. It will therefore always be less accurate than therapeutic drug monitoring. For genotyping, rare mutations, effects of concomitant medications, diet, or underlying kidney or liver problems are not included in this prognosis. Yet, it is believed that, especially with the aim of reducing costs, it may be beneficiary to have the genetic information on the most important enzymes in drug metabolism analysed pre-emptively. In the Netherlands, patients can receive a "DNA-passport for Medication", with which they can get advice on drug dosing at each Pharmacy in the Netherlands (Fig. 7.11 and Table 7.2).

Table 7.2 Pharmacogenetic markers

Gene	Drug	Intended use
BChE	Succinylcholine	Muscle relaxant by surgery
CYP2B6	efavirenz	HIV
CYP2C9/VKORC1	Acenocoumarol	Anticoagulation
	Phenprocoumon	Anticoagulation
	Warfarin	Anticoagulation
CYP2C19	Clopidogrel	Inhibitor platelet aggregation
CYP2D6	Tricyclic antidepressives	Depression
	Antipsychotic	Psychoses
	Tamoxifen	Breast cancer
CYP3A5	Tacrolimus	Organ transplantation
DPYD	5-FU, capacetabin	Colon cancer
HLA-B*5701	Abacavir	HIV
IL28B	Inteferon/ribavirin	Hepatitis C
TPMT	Azathioprine	Dermatology, rheumatology, Crohn's disease
UGT1A1	Irinotecan	Colon cancer

7.5 Haemoglobinopathy

Each red blood cell contains hundreds of millions of copies of the protein haemoglobin (Hb). The main function of haemoglobin is to transport oxygen and each haemoglobin molecule is built up of four protein chains. The most common types of haemoglobin are HbA0, HbA2 and HbF. In adults, HbA0 forms the largest fraction, with approximately 96–98% of the total Hb, followed by HbA2 (2.2–3.2%), and HbF (<2%). HbA0 is composed of two α-globin chains and two β-globin chains, HbA2 consist of two α- and two δ-globin chains, and HbF is made up of two α-globin chains and two γ-globin chains. The α-globin chains consist of 141 amino acids that are encoded by two closely linked genes (α1 and α2) located on chromosome 16. The β-globin chains consist of 146 amino acids that are encoded by the β-gene, which is located on chromosome 11 (Fig. 7.12). Both α- and β-globin chain encoding genes contain three exons and two introns.

Mutations in the genes encoding for the protein chains of haemoglobin can lead to haemoglobinopathies. Haemoglobinopathies are divided into two different classes: thalassemia's and haemoglobin variants. In the case of α- and β-thalassemia's, fewer α- or β-globin chains are being produced than normal, implicating a quantitative defect. On the other hand, haemoglobin variants are usually haemoglobin proteins with substituted amino acids at particular loci in the α-globin or β-globin chain.

The symptoms of haemoglobinopathies range from asymptomatic (no symptoms at all, just abnormal laboratory findings) to severe, wherein patients are dependent

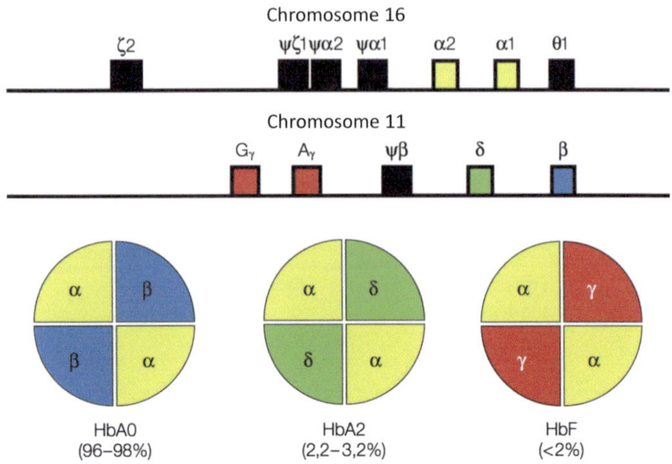

Fig. 7.12 Schematic representation of the α-en β-globinegene clusters. 'ψ' is a pseudogene. Combination of the gene products is indicated, with the nomenclature and frequencies

on blood transfusions at regular intervals, or which are not even compatible with life. In fact, during blood transfusions, patients are administered red blood cells of healthy donors to substitute for their own defective or deficient haemoglobin.

Ever since their introduction and further development, molecular diagnostic technologies have played a crucial role in the diagnosis of haemoglobinopathies. Nowadays, many clinical chemistry laboratories can perform molecular diagnosis of haemoglobinopathies.

Diagnosing haemoglobinopathies is important for at least three reasons:

- For confirmation of clinically significant thalassemia's or of other genetic haemoglobin disorders
- For identification of carriers of a haemoglobinopathy, which is important to reduce unnecessary diagnostics or therapy as for example, laboratory findings and blood film morphology can easily be mistaken for anaemia caused by iron deficiency. The treatment of iron deficiency anaemia with iron is risky, as iron overload is possible in patients with thalassemia's, due to the potential excessive red cell destruction in haemoglobinopathies. Thus, treatment can deteriorate an existing iron overload, which can eventually lead to organ damage, as there is no physiological mechanism for eliminating excess iron from the body
- Carriers of a haemoglobinopathy have an increased risk of severe forms of haemoglobinopathies in their offspring. So, it is important to anticipate this by performing molecular diagnostics, followed by referral for counselling at a specialized clinical genetics centre.

Thalassemias are generally characterized by an abnormal blood image, including microcytic anaemia [reduced Hb and reduced MCV (Mean Cell Volume)] and an increased number of erythrocytes in relation to the amount of haemoglobin. If a

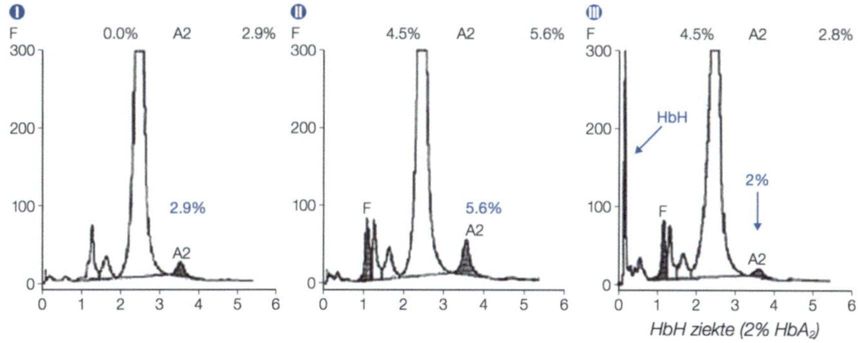

Fig. 7.13 HPLC chromatogram, showing the HbA2, HbF and HbH peaks

genetic disorder of haemoglobin is suspected, the next step is analysis of the most common types of haemoglobin in the blood of the patient. This is usually done with the aid of capillary protein electrophoresis, or High Performance Liquid Chromatography (HPLC) (Fig. 7.13). The HbA2 fraction in a HPLC run is characteristically normal or reduced in carriers or in clinically affected α-thalassemia patients, but mostly increased in β-thalassemia. In both situations, confirmation of the disease diagnosis can only be made by molecular diagnostics and this is also required to determine the exact type of the haemoglobin abnormality.

As mentioned earlier, normally, four functional α-globin chain-encoding genes are present. The used nomenclature is: (αα/αα). The absence of four functional α-globin genes (– –/– –) is not compatible with life, and children born with this defect die at birth due to fetal hydrops. A second severe form of α-thalassemia is known as Hb H disease, wherein just one active α-globin gene is present (–α/– –), and affected patients depend on blood transfusions. In Hb H disease, the β-chains cannot associate in a stoichiometric ratio with α-chains and thus complexes of excess β-chains are formed, which can be detected in the red blood cells of such patients. The reverse effect can be seen in β-thalassemia. Generally, the greater the excess of unassociated α- or β-chains, the more severe the reduction of Hb in blood and the worse the anaemia is. Forms of α-thalassemia with one or two α-globin genes missing (–α/αα, –α/–α and – –/αα), generally have a mild course without significant clinical symptoms.

The majority of α-thalassemia (80%) is caused by large deletions of DNA, resulting in parts wherein one or more α-globin genes are lost. The remaining 20% are caused by point mutations. The most occurring deletional α-thalassemia's are: $-\alpha^{3.7}$, $-\alpha^{4.2}$, $-(\alpha)^{20.5}$, $--^{SEA}$, $--^{MED}$, $--^{THAI}$ and $--^{FIL}$. These types of deletions are routinely diagnosed in clinical chemistry laboratories by gap-PCR techniques. In gap-PCR, primers are designed in a way that a particular PCR product can only be formed if there is a deletion in the DNA target defined by the primer design. Without such a gap, no PCR product can be formed, as the distance between the primers is too large. Conversely, the gap-PCR can also be designed in such a way

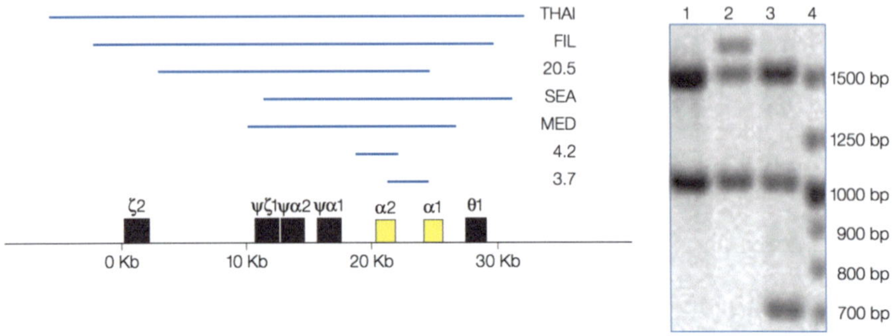

Fig. 7.14 Schematic representation of the 7 most common deletions in the α-globine gene cluster, resulting in α-thalassemia: THAI = Thailand, FIL = Filipino, 20.5 + −(α)20.5, SEA = South-East Asia, MED = Mediterranean, 4.2 = -α4.2, 3.7: −α3.7

that, if there is a gap due to a deletional thalassemia, shorter PCR products are formed instead of no PCR products. Figure 7.14 illustrates the application of gap-PCR for the diagnosis of α-thalassemia.

In contrast to α-thalassemias, most β-thalassemias (>95%) are caused by point mutations rather than gene deletions. Sequencing of the β-chain is the best option for molecular diagnostics. Until now, a few hundred point mutations are known which produce β-thalassemia. In this context, it is useful to know that severe β-thalassemia (i.e. thalassemia major) often results in cases wherein two different point mutations are inherited. If no point mutations can be found, possible existence of deletions of (parts of) the β-gene can be excluded with gap-PCR or MLPA (see Vol. 1, Sect. 5.12.8). Patients with homozygous or compound heterozygous β-thalassemia may have insufficient oxygen transport to the tissues and may need blood transfusions.

Dedicated clinical chemistry and haematology laboratories using more specialized techniques, such as Sanger sequencing, have the capacity to sequence globin genes in order to investigate the presence of point mutations. The globin genes are amplified by PCR and the resulting PCR products are sequenced using the dideoxy sequencing method (see Vol. 1, Chap. 8). More rare thalassemias caused by DNA deletions can be detected more efficiently by multiplex ligation-dependent probe amplification (MLPA) (see Vol.1, Sect. 5.12.8). The MLPA technique cannot be applied by most laboratories yet, but can be carried out at reference laboratories for research and diagnosis of haemoglobinopathies. In short, MLPA makes use of a mixture of MLPA probes (45–60 probes) consisting of two different hemi-probes that must hybridize to immediately adjacent target sequences in order to be ligated onto a single probe (see Vol. 1, Fig. 5.33). However, if there is a deletion or a mutation, a single specific probe is not able to hybridize with its counterpart in a perfect tandem way, or not at all. Eventually, all probes are amplified simultaneously using the same PCR primer pair. As all probes have different lengths, a deletion can be easily visualised, as it is not or just partly amplified.

Virtually all currently known DNA variants are caused by point mutations. Carriers of a haemoglobin variant are confirmed by DNA sequencing of the β-globin genes, or, in rare cases, the α-globin genes. In the majority of cases of the hundreds of known haemoglobin variants that have been identified to date, only a small numbers of variants are clinically relevant.

Probably the most well-known and clinically most relevant haemoglobinopathy is sickle cell anaemia. In The Netherlands, around 40–60 children are born each year with homozygous sickle cell anaemia. A single base change, being the substitution of adenine by thymine (GAG to GTG in codon 6), encodes an amino acid change at the sixth position of the β-chain. Whereas a homozygous sickle cell type HbSS causes a severe variant of haemoglobinopathy, carriers of HbS (heterozygous sickle cell anaemia, type HbAS) are in fact more protected against malaria, which keeps this trait present in malaria endemic areas. Sickle cell anaemia disease is associated with severe pain during sickle cell crises and is a medical emergency. These crises occur when red blood cells deform from their normal biconcave shape into crescent-shaped sickle cells under the influence of low oxygen tension. As a result, small blood vessels are occluded, hereby creating a shortage of oxygen in tissues and organs, and thus severe pain.

7.6 Molecular Diagnostics in Hematologic Malignancies

Pathophysiology

The production of our blood cells primarily occurs in the bone marrow. Three main types of blood cells (erythrocytes, leukocytes and platelets) emerge from the (bone marrow) stem cells by multiplication and maturation. DNA damage in a stem cell may interfere with the proliferation and differentiation and may result in the accumulation of abnormal blood cells. When mainly malignant cells proliferate and accumulate in the bone marrow and blood, it is classified as leukaemia. If these cells are primarily confined to lymphoid tissue, such as the lymph nodes, this is called a lymphoma. Leukaemias are divided in acute and chronic diseases. The acute and chronic leukaemias are, depending on the cell type, divided into lymphoid (B and T lymphocytes) and myeloid (erythrocytes, platelets, granulocytes, and monocytes) types (see also Chap. 6).

Indicator Targets for DNA diagnostics

DNA diagnostics of haematological malignancies has three indications:

1. Confirmation of a malignancy
2. Classification of a malignancy
3. Monitoring.

The first is to confirm a malignancy. This is achieved by proving the presence of a monoclonal lymphocytic cell population. Cells of a malignancy are of clonal origin: they are the result of one degenerated cell.

In malignant lymphocytic leukaemia's and lymphomas (75% of all hematologic malignancies), over 98% of patients have clonally rearranged immunoglobulin (Ig) and T-cell receptor (TCR) genes (the molecular-biological backgrounds of this are described in paragraphs 7.1, 7.2 and 7.3). This is in contrast with the normal situation, in which there is a great diversity of Ig/TCR gene rearrangements (polyclonality), in order to guarantee a broad immunity against foreign antigens.

Chromosomal translocations can be used as a marker for clonality. Here, parts of two genes, usually located on different chromosomes, fuse on one chromosome in a tandem way, thereby coding for an aberrant (fusion) protein. If the product is subsequently involved in cell proliferation and/or maturation, it could lead to deregulation and a malignant transformation. The best-known example of such a chromosomal abnormality is the Philadelphia chromosome, a fusion of BCR/ABL, which occurs in almost all patients with chronic myeloid leukaemia (CML) (see also Fig. 6.8).

The frequency of molecularly detectable chromosome aberrations varies greatly from one disease category to the other. It is less than 5% in chronic lymphocytic leukaemias to about 35% in case of acute leukaemias. In lymphomas it increases to 70–80% and chronic myeloid leukaemias show levels of 95% or more.

Another indication for DNA testing involves the classification of haematological malignancies, which were formerly mainly based on morphological and immunological characteristics. Nowadays, the identification of specific chromosome aberrations using molecular techniques contributes to the diagnosis and classification of a part of the malignancies. Herewith, differentiation in treatment strategies comes into the therapeutic arsenal. In the meantime, molecular characterization was able to distinguish apparently identical forms of leukaemia into good and poor prognosis subtypes. An example of this is promyelocytic leukaemia (PML), with a subtype-specific translocation between chromosomes 15 and 17 (t (15; 17)). The resulting fusion protein has an altered vitamin-A-receptor; thereby keeping cells in growth arrest at a specific stage of development. When patients with this form of leukaemia are treated with high doses of vitamin A, the leukemic cells do mature, leaving the leukemic state. Such patients therefore don't need to undergo intensive chemotherapy or a bone marrow transplant. In these cases vitamin A therapy might be sufficient, possibly supported by low-dose chemotherapy.

A third and last indication for DNA diagnostics concerns the monitoring of malignancy during treatment. Most current patients with leukaemia or lymphoma reach complete remission (= disappearance of disease) with the current treatment protocols (chemotherapy). Bone marrow transplantation also appears to be able to contribute to the prolongation of remission and survival. However, relapse (= return of the malignancy) does occur in some patients. Apparently, the treatment protocols in these patients have not been able to eradicate all malignant cells, even though patients were in complete remission according to traditional and clinical criteria.

In order to gain sufficient insight into the effectiveness of the treatment, sensitive techniques are needed to detect very small numbers of malignant cells, or "minimal residual disease" (MRD). If tumour-specific DNA or RNA sequences, such as fracture point-spanning regions of chromosome aberrations are present, they can be used as PCR targets. The required sensitivity of 10^4 to 10^5 (that is to say, a single leukaemia cell is detectable within a population of 10,000 to 100,000 normal cells) can be reached. MRD-detection during the first three to four months of treatment identifies patients with good or poor responses to therapy accurately. Both the rate of disappearance of leukemic cells after induction therapy and the absolute number of remaining leukemic cells appear to be strong indicators of whether or not recurrence of disease will occur (van Dongen et al. 1998).

Techniques

For laboratory diagnosis of haematological malignancies, cytomorphological and immunological techniques are essential. In addition, Southern blotting, PCR and FISH are important diagnostic tools.

Rearrangement of Ig and TCR genes can be detected at the DNA level with Southern blotting and by PCR analysis (see Vol. 1, Chap. 4). Chromosomal translocations can be visualised by cytogenetic techniques (karyotyping). Karyotyping requires fresh material and living (cultured) cells. Only chromosomal abnormalities for which tumour specific probes are available and that reach the detection limit of the fluorescence microscope can be detected. Due to this, the applicability of the technique is limited. Alternatively, chromosomal translocations can be demonstrated at the DNA level with Southern blotting and PCR and at the RNA level by RT-PCR. The sensitivity of PCR is a factor of 100–1000 higher compared to Southern blot analysis.

It is important to include cytological features in the interpretation of the results of cytogenetic [analysis] and molecular genetic analyses, because identical chromosomal abnormalities can be present in various forms of leukaemia. Furthermore, different laboratories have to match their techniques in order to reach comparable sensitivity and interpretation. This was achieved recently by establishing a molecular diagnostics network with the aim to implement new molecular diagnostic methods in a uniform manner in the clinic within the Foundation HOVON (Hemato Oncology Adults Netherlands) and the Dutch Society for Haematology (NVvH). The network is initially focused on the molecular diagnosis of leukaemia.

7.7 Molecular Diagnosis of Rare Genetic Disorders

Specific research to diagnose a rare genetic abnormality is usually not performed at peripheral clinical chemistry laboratories, but only in academic centres. However, recent technological developments do bring this type of diagnostics within the reach of clinical chemistry laboratories of at least several large, non-academic centres as well. The diagnosis of a hereditary disease often implies a large impact on medical and social aspects for patients. Therefore it requires clinical genetic counselling to support and to educate the patients and their relatives. However, a first screening round can take place at Clinical Chemistry laboratories, such as for Cystic Fibrosis, to keep the burden of work for Clinical Genetic centres manageable. In such a screening, the most frequent mutations are analysed. These laboratories work closely together with Clinical Genetic centres for further referral of patients if indicated and for monitoring of the quality of DNA analysis.

7.8 Molecular Diagnostics of Solid Tumours

It has been known since 1948 that cell free DNA can be found in plasma of healthy individuals (Mandel 1948). In plasma of patients with cancer, higher concentrations of this type of DNA are present. It is easy to understand why: in a tumour, there is a lot of cell proliferation, and also increased necrosis and apoptosis. As a result of these processes, DNA from the tumour will be released into the circulation. This also applies to autoimmune diseases, trauma, myocardial infarction, etc. Within this circulating DNA, it is possible to identify tumour-specific mutations. This is important, since knowledge of the mutations driving a tumour is essential for effective treatment. The ability to measure tumour-derived mutations in blood gives a wide range of options to optimize the diagnosis of cancer and its treatment for the individual patient. Circulating tumour DNA could for example allow tumour diagnostics in the absence of a tissue result, be used for frequent follow up, screening for emerging resistance, detection of recurrence, etc. Since the analysis of ctDNA is based on material obtained by a venipuncture, the impact on the individual patient is limited when compared to tissue biopsies (which is at this moment required for mutation analysis). An additional important difference with tissue biopsies is the fact that the cell free DNA can reflect mutations from the primary tumour (including the presence of multiple clones), but also from metastatic lesions. Cell free DNA seems to reflect the total tumour heterogeneity better than a tissue biopsy.

So, although the tumour is not present in the blood, its DNA can be identified in the circulation as cell free DNA. Specific tumour-derived mutations distinguish the DNA derived from the tumour from the somatic DNA. For example, a deletion in the "Epidermal Growth Factor Receptor" (EGFR) leads to an abnormal activation of the RAS-RAF-MEK-MAPK pathway in the cell. This in turn causes increased cell proliferation, prevention of apoptosis, an increase in protein production, etc. Eventually, this leads to the survival of aberrant cells and an uncontrolled cell growth. As we already mentioned, knowledge of these changes is essential since different mutations or combinations of mutations require different treatment. For example, an EGFR mutation can be treated with erlotinib. Erlotinib blocks the activation domain of the EGFR. This suppresses the RAS-RAF-MEK-MAPK pathway and the effects of the route will be cut off. This therapy works well, but its effectiveness is limited. New mutations can occur or undetected resistant tumour cells can be present before treatment.

There are several well-known resistance mechanisms. In approximately 50% of the patients treated with erlotinib or gefitinib, the EGFR T790M mutation can be identified. This mutation results in a conformational change, causing agents such as erlotinib to no longer to be able to bind to the EGFR receptor. Other mechanisms of resistance can include mutations in KRAS, BRAF, PIK3CA, EGFR or Her2 amplifications. In 15–20% of the patients, the resistance mechanisms are unknown (Sequist et al. 2011). In this context, cell free DNA could allow early identification of emerging mechanisms of resistance and thereby select the most effective therapy at the right time for the right patient, but also limit the time in which ineffective (expensive) treatment is continued.

Analysis of cell free DNA is technically demanding, since the tumour-derived DNA forms a small subpopulation in a background of wild type DNA. This poses specific technical challenges. Recent technical developments make it possible to identify tumour-specific mutations in cell free DNA (cf DNA) in the circulation (blood) of cancer patients. One of these techniques is especially suited and is discussed below.

Digital PCR

A fairly recent development within PCR technology is digital PCR (Vogelstein 1999).

Digital PCR differs from qPCR in two aspects. First, the PCR mix (target DNA and PCR mix) is distributed over a large number of compartments in digital PCR. This can be achieved by single miniature droplets (20,000–10,000) (Fig. 7.15) or in the form of a chip. By distributing a sample with a certain amount of DNA over a

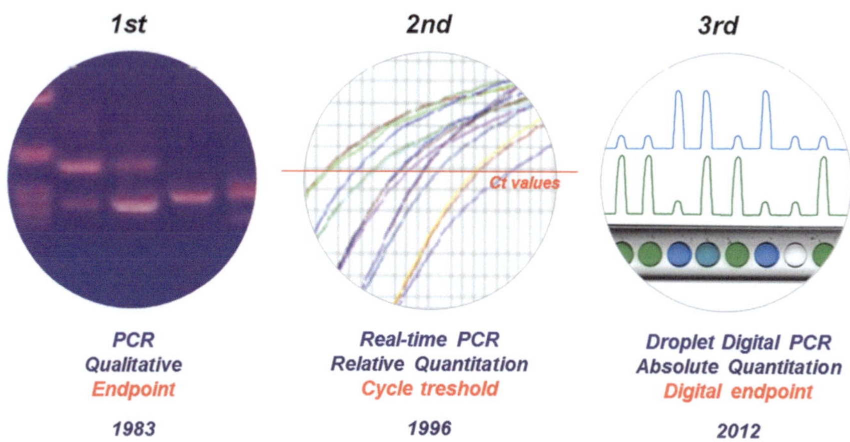

Fig. 7.15 Evolution of PCR, going from gel based endpoint analysis (qualitative endpoint), through real-time PCR (quantitative by cycle threshold measurements) to droplet digital PCR, using (positive) droplet counting as a digital quantitative endpoint

large number of droplets or compartments, the DNA fragments will be normally distributed over the droplets. As a result only one (or a few) DNA fragment will be present in a single droplet or compartment. In these droplets, a qPCR reaction is carried out. Specific hydrolysis probes for wild type and mutated sequences identify individual DNA molecules. As a result, a fluorescent signal will be present in the droplet when a DNA fragment is amplified. By using different fluorescent markers, droplets containing a DNA fragment with a mutation can easily be discriminated from the other droplets that contain the wt-DNA. A second difference is that digital PCR is an end-point PCR reaction. This means that individual droplets will be identified based on the end product of a number of PCR cycles and not on a Cq value. By counting "positive" droplets or compartments, allele frequencies of 0.1–0.001% of mutated DNA can be detected. This means that digital PCR is a highly sensitive method to identify low frequency DNA fragments in a large background population of wild type DNA fragments. Clinical applications are for instance the detection of Epidermal Growth Factor Receptor T790M mutation, which causes resistance to gefitinib therapy in lung cancer. As soon as this mutation appears in the tumour, therapy needs to be adjusted. Currently, this analysis is mostly made on tumour biopsy material, but by using digital PCR, a far less invasive blood sample can be investigated with an even higher sensitivity than a biopsy (Fig. 7.16).

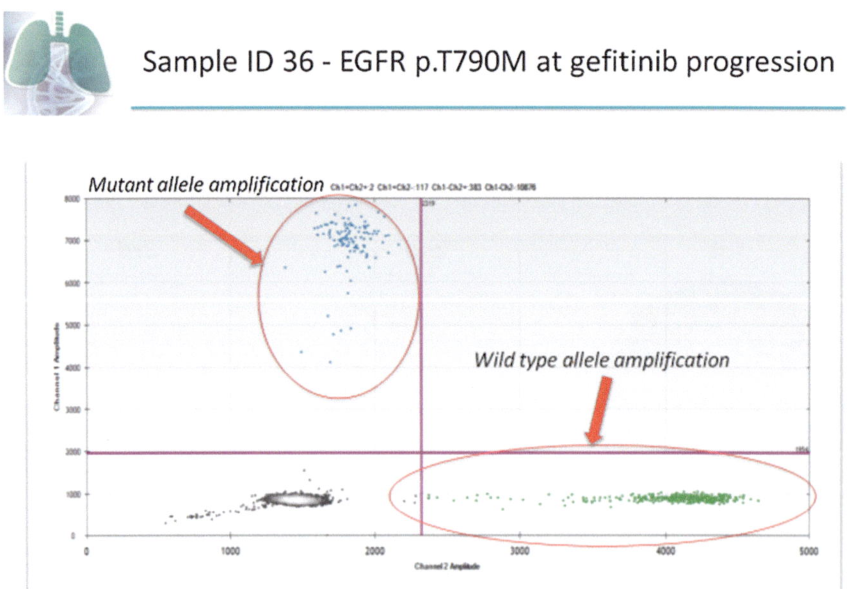

Fig. 7.16 Output of a droplet digital PCR experiment for EGFR p.T790M resistance mutation, obtained from circulating cell free DNA from plasma of a lung cancer patients. The EGFR p. T790M positive droplets in blue indicate positivity for this mutation, with the number of droplets being directly proportional to the amount of EGFR p.T790M positive DNA molecules. In *green*, the EGFR wild type signal is seen, enabling calculation of a p.T790M/EGFR wild type ratio

One of the disadvantages of PCR based techniques is that only a limited number of mutations can be analysed using droplet digital PCR. Due to the complexity and the evolution of solid tumours in the course of the disease, there is a need for techniques combining a high sensitivity with the possibility to analyse multiple targets. Several methods have been published, for example, BEAMing, barcoding, Tam-seq, Capp-seq, PARE and whole exome sequencing. A definitive technical solution and a standardisation of pre-analytical and analytical procedures are, however, not yet available.

7.9 The Future

Developments in molecular diagnostics are increasingly focused on sequencing the entire genome (whole genome sequencing, WGS). Using WGS, it will no longer be an issue which polymorphisms or mutations need to be searched for by specific techniques such as PCR-RFLP, hybridization techniques, Sanger sequencing, or other approaches: the DNA sequence will already be available. Bioinformatic analyses can be done to find the variants or mutations of interest. In particular in the field of inherited diseases and genetic variations this will most probably be the approach that will be followed. For oncology, sequencing should be done on the tumour: which is constantly changing in terms of DNA sequence. DNA analyses focus on characterization, tumour load and/or treatment options. The coming years will be important in determining whether whole genome sequencing, in terms of reliability and cost, can be applied routinely in patient diagnostics.

Further Reading and References

Aillaud MF, et al. Resistance to activated protein C-diagnostic strategy in a laboratory of haemostasis. Tromb Haemost. 1995;74:1197–207.

Bertina RM, et al. Mutation in blood coagulation factor V associated with resistance to activated protein C. Nature. 1994;369:64–7.

European Association For The Study Of The Liver. EASL clinical practice guidelines for HFE hemochromatosis. J Hepatol. 2010 Jul;53(1):3–22.

Evans DAP, et al. Genetic control of isoniazid metabolism in man. Br Med J. 1960;2:482.

Evans DAP. An improved and simplified method of detecting the acetylator phenotype. J Med Genet. 1969;6:405–7.

Feder JN, et al. A novel MHC class I-like gene is mutated in patients with hereditary haemochromatosis. Nat Genet. 1996;13:399–408.

Feder JN, et al. The hemochromatosis founder mutation in HLA-H disrupts beta2-microglobulin interaction and cell surface expression. J Biol Chem. 1997;272:14025–8.

Grandone E, et al. Factor V Leiden, C.T MTHFR polymorphism and genetic susceptibility to pre-eclampsia. Thromb Haemost. 1997;77:1052–4.

Koeleman BPC, et al. Activated protein C resistance as an additional risk factor for thrombosis in protein C-deficient families. Blood. 1994;84:1031–5.

Mandel P, et al. Les Acides Nucleiques de plasma sanquin chez l'homme. C.R. Seances Soc Biol Fil 1948 Feb;142(3–4):241.
Moderne Vogel F, der Humangenetik Probleme. Ergebn Inn Med Kinderheilkd. 1959;12:52–125.
Nederlandse Richtlijn voor de Diagnostiek en behandeling van hereditaire hemochromatose 2007. http://www.internisten.nl/uploads/AL/FJ/ALFJChidQ2OgF7uax8yh1w/richtlijn_2007_Hemochromatose.pdf.
Porto G, et al. EMQN best practice guidelines for the molecular genetic diagnosis of hereditary hemochromatosis (HH). Eur J Hum Genet. 2016 Jul 8;24:479–95. doi:10.1038/ejhg.2015.128.
Schaeffeler E, et al. Comprehensive analysis of thiopurine S-methyltransferase phenotype-genotype correlation in a large population of German-Caucasians and identification of novel TPMT variants. Pharmacogenetics. 2004 Jul;14(7):407–17.
Sequist LV, et al. Genotypic and histological evolution of lung cancers acquiring resistance to EGFR inhibitors. Sci Transl Med. 2011;3(75).
Sim SC, et al. A common novel CYP2C19 gene variant causes ultrarapid drug metabolism relevant for the drug response to proton pump inhibitors and antidepressants Clin Pharmacol Ther. 2006 Jan;79(1):103–13.
Swen JJ, et al. Pharmacogenetics: from bench to byte–an update of guidelines. Clin Pharmacol Ther. 2011 May;89(5):662–73. doi:10.1038/clpt.2011.34. Epub 2011 Mar 16.
van Dongen JJ, et al. Prognostic value of minimal residual disease in acute lymphoblastic leukaemia in childhood. Lancet. 1998;352:1731–8.
van Bokhoven MA, et al. Diagnosis and management of hereditary haemochromatosis. BMJ. 2011 Jan 19;342:c7251.
Vogelstein B, et al. Digital PCR. Proc Natl Acad Sci. USA 1999;96:9236–41.
Weinshilboum RM, Sladek SL. Mercaptopurine pharmacogenetics: monogenic inheritance of erythrocyte thiopurine methyltransferase activity. Am J Hum Genet. 1980 Sept;32(5):651–62.

Chapter 8
Clinical Genetics

A.M.W. van den Ouweland, R. van Minkelen, J. Knijnenburg, M. van Slegtenhorst and L.H. Hoefsloot

Abstract Clinical Genetics is the field studying the inheritance and causes of genetic disorders. A genetic disorder is defined as a pathological condition caused by an absent or defective gene or by a chromosomal aberration. Knowing the cause of a genetic disorder can be of crucial importance, not only for the patient, but also for his or her family. This chapter describes the use of several molecular methods, using examples from daily practice in the clinical genetic laboratory. The use of SNPs (single nucleotide polymorphisms) to detect large deletions and duplications is discussed. One paragraph describes how single gene sequencing is used to detect small mutations (nucleotide changes, small deletions and/or insertions) in the *BRCA1* and *BRCA2* genes that are tested routinely in the laboratory and take up more than 20% of the tests performed. The diagnosis of patients with a repeat expansion as their causative mutation asks for a different approach. Finally, the diagnostics for patients with intellectual disability is described. In the past years these patients often had no diagnosis, but with the introduction of exome sequencing in clinical practice, up to 30% of these patients now have a definitive diagnosis.

Keywords Molecular diagnostics for inherited disease · Clinical genetic laboratory · Diagnostic microarrays · Clinical whole exome sequencing

8.1 Introduction

Clinical Genetics is the field that studies the inheritance and causes of genetic disorders. Genetic disorders can be very serious; it is usually defined as a pathological condition caused by an absent or defective gene, but also chromosomal aberrations involving much larger DNA rearrangements can be causative of a

A.M.W. van den Ouweland · R. van Minkelen · J. Knijnenburg ·
M.van Slegtenhorst · L.H. Hoefsloot (✉)
Clinical Genetics, Erasmus MC, Rotterdam, The Netherlands
e-mail: e.hoefsloot@erasmusmc.nl

E. van Pelt-Verkuil et al. (eds.), *Molecular Diagnostics*,
DOI 10.1007/978-981-10-4511-0_8

genetic disorder. Examples of the latter are for instance Down syndrome patients, who have an extra copy of chromosome 21. There are different methods to identify a genetic disorder. Historically, the field was divided in metabolic, enzymatic, cytogenetic and DNA diagnostics laboratories. Depending on the clinical presentation and the question of the patient, the method of choice could involve several of these laboratories. For instance, a patient that was suspected to have a lysosomal storage disorder would first be sent into test the lysosomal enzymes in the enzyme laboratory, and only when this was shown to be defective would the appropriate gene be sequenced in the DNA diagnostic laboratory to identify the causative mutation. However, with the introduction of massive parallel sequencing (next generation sequencing) the field of genetic diagnostics has been changing more rapidly than ever before.

In the Netherlands, only the academic medical centres (AMC, VUMC, LUMC, Erasmus MC, UMCU, UMCG, Radboud MC and MUMC) have permission from the government to offer DNA diagnostics of hereditary disorders. The one exception is the NKI-AVL (Dutch Cancer Institute), which specializes in the counselling and DNA diagnostics of hereditary tumours.

8.2 General

The human DNA contains 3 billion base pairs, divided over 23 chromosome pairs. The first 22 chromosomes are called autosomes, the 23rd pair are the sex chromosomes, XX for females and XY for males. Basically, you inherit one chromosome set from your father, and one chromosome set from your mother. There are several modes of inheritance for genetic disorders: autosomal recessive, autosomal dominant, X-linked and mitochondrial (see Table 8.1 and Vol. 1, Chap. 2).

Autosomal recessive inheritance is the mode of inheritance in which both copies of a gene (located on one of the autosomes) need to be defective in order to cause clinical symptoms. Typically, both parents are carriers without symptoms, and the children have a 1:4 risk to inherit both defective copies of their parents and become ill.

Table 8.1 Modes of inheritance

Mode of inheritance	DNA level	Characteristics of families
Autosomal recessive	Autosome: both gene copies affected	Sporadic/one generation, only brothers and sisters affected in 1:4 ratio, parents not affected
Autosomal dominant	Autosome: one gene copy affected	All generations affected in 1:2 ratio, male to male transmission; sporadic patients when *de novo*
X-linked	Gene copy located on X-chromosome	Females not affected (1:2 are carrier), males affected (1:2), no male to male transmission
Mitochondrial	Gene located on mitochondrial DNA	Variable expression, males and females equally affected, no male to male transmission

Autosomal dominant inheritance is seen in families in which only one defective copy is enough to cause disease. This means that there are affected family members in more than one generation and also, that the chance for the offspring of an affected person to be affected as well is 1:2. In several disorders with an autosomal dominant inheritance, only sporadic (single affected in the family) patients are described; and this can resemble autosomal recessive inheritance. In these disorders a de novo mutation is identified, i.e. the mutation has arisen in the patient and has not been inherited from the father or the mother. In these cases, it cannot be excluded that one of the parents is germline mosaic for the mutation. Many patients with intellectual disability have a de novo mutation in a gene that is expressed in the brain. The one healthy copy of the gene is not enough to code for the concentration of protein needed in certain stages of development (haploinsufficiency), and this leads to an aberrant development. Alternatively, the mutant copy of the gene codes for a protein that has a different (deleterious) effect compared to the normal function of the protein (gain-of-function mutation).

In X-linked inheritance, the gene is located on the X-chromosome. Females that are carriers of a defective gene copy on one of their X-chromosomes are not affected, but males that inherit a defective gene copy on the one X-chromosome they inherit from their mother are clinically affected (for instance Duchenne muscular dystrophy).

Finally, mitochondria have their own DNA, and mitochondria are inherited from the mother as sperm does not contribute mitochondria to the zygote. Typically, the disease phenotype associated with a mitochondrial mutation is very variable, but daughters and sons can both be affected. For further in-depth information please refer to Turnpenny and Ellard 2011.

Finding the causative mutation in the DNA of a patient is the core business of the clinical genetic laboratory. Several technologies can be applied to identify structural changes in the DNA that underlie genetic disorders. The decision to use a certain technology is basically dictated by the size and type of the structural change that is expected. When looking at large rearrangements, like chromosomal abnormalities, karyotyping has long been the golden standard. With karyotyping, whole chromosomes can be visualised under the microscope using dividing cells and specialised colouring (banding) techniques. In recent years, genome wide molecular diagnostic screening using microarrays for copy number alterations has become the method of choice for most applications. This type of screening has proven to be instrumental in retrieving the genetic cause of e.g. ultrasound abnormalities and intellectual disability with or without congenital malformations in a significant portion of patients. Routine diagnostics for this type of patients is nowadays performed using microarrays, allowing detection of copy-number imbalances down to approximately 100 kb. Next to the detection of congenital genetic alterations, this technique is also increasingly used in the detection of somatic genetic alterations that may play a role in the diagnosis and prognosis of malignancies. This is discussed in Sect. 8.3. For smaller deletions and duplications, MLPA can be used (see Vol. 1, Chap. 5 for an explanation about this technique).

Point mutations, rearrangements (deletions/duplications) and repeat expansions can also be the cause of several disorders. Repeats of nucleotides (e.g. a $(CGG)_n$ repeat) are present in all individuals. The number of nucleotides (a triplet in this example) that is repeated can differ for each individual in the same way people can have different alleles for a single nucleotide polymorphism (SNP), i.e. it is part of the natural variation of the genome. In the early 1990s is was found that these so-called inherited trinucleotide repeat (TNR) expansions can however expand beyond a ‘normal’ range and become pathogenic. These expanded repeats are pathogenic because they affect the expression of a gene and therefore also influence the function of the protein encoded by that gene. This can be either by a loss of function mechanism (for example Fragile X Syndrome) or a gain of function mechanism (for example Myotonic Dystrophy). For some disorders, the pathogenic mechanism is not known yet (for example Spinocerebellar ataxia type 12).

The disorder Fragile X Syndrome is used in Sect. 8.4 as example how these repeat expansions can be determined.

Many indirect screening techniques have been used in the past for the detection of single nucleotide changes and small deletions and insertions. The choice for a certain method is made on the basis of sensitivity, robustness, hands-on time, and costs. These indirect screening methods include the Protein Truncation Test (PTT), High Resolution Melting curve analysis (HRM), and Denaturing Gradient Gel Electrophoresis (DGGE). When a PCR product showed an aberrant pattern in one of these tests, this PCR product needed to be sequenced to know the exact nucleotide change. Due to the availability of robust robots for all pipetting work, reduced costs of sequence kits and automated sequence analysis programs like SeqPilot (JSI Medical Systems), SeqScape (Life technologies) and Mutation Surveyor (Softgenetics) it became possible to directly sequence individual genes in a diagnostic setting. The use of Sanger sequencing will be demonstrated in the diagnostics for families with breast/ovarian cancer (Sect. 8.5). The technical details of Sanger sequencing are given in Vol. 1, Chap. 8.

The introduction of massive parallel sequencing (next generation sequencing, also discussed in Vol. 1, Chap. 8) has changed the diagnosis of genetic disorders into a completely different ball game. Some genetic disorders are extremely heterogeneous, meaning that the same condition can be caused by a mutation in many different genes. For intellectual disability >1000 genes have now been described, but also for conditions like loss of vision (retinitis pigmentosa and disorders of the macula) more than 100 genes have been described. Finding the causative mutation for these patients was virtually impossible in the past, because Sanger sequencing of all known genes sequentially and individually took too long and was too expensive. Nowadays, using whole exome sequencing, followed by a filter for the variants that appear in the genes associated with the disorder is the method of choice in genome diagnostics (see Sect. 8.6).

In Table 8.2, an overview of differently sized mutations and their consequences are given, together with the preferred method of detection.

Table 8.2 DNA mutations and their consequences

	Size	Coding region: consequence	Non-coding region: consequence	Detection method
Single nucleotide change	1 bp	• Amino acid change • Stop codon • Splice site destruction	• Splice site destruction • Changes in transcription and/or translation	Sequence analysis (Sanger or massive parallel)
Small deletions/insertions	1–30 bp	• In frame deletion or insertion of amino acids • Frame shift, leading to premature stop	• Changes in transcription and/or translation	Sequence analysis (Sanger or massive parallel)
Trinucleotide expansions	30–2000 bp	Expansion of poly-amino acid stretches	Disruption of splicing machinery	Triplet PCR
Larger deletions	<1 kb and larger	Partial or complete gene deletions	Changes in transcription and/or translation	MLPA, SNP array, (Southern blot)
Larger insertions	<1 kb and larger	Disruption of coding region	Duplication of dosage sensitive gene	MLPA, SNP array, (Southern blot)
Chromosomal aberrations	Partial or whole chromosomes	Chromosomal syndrome	Chromosomal syndrome	SNP array, karyotyping

8.3 Genomic Microarrays

Since the discovery of the extra chromosome 21 in Down patients, the analysis of larger structural DNA changes is performed diagnostically. The techniques are used to identify causes for intellectual disability (ID) and/or multiple congenital anomalies (MCA), ultrasound anomalies (US) or to identify aberrations of prognostic significance in haematological malignancies. Until ca. 10 years ago, classical chromosome banding techniques (karyotyping), supplemented with fluorescence in situ hybridisation (FISH) for detection of specific microdeletions/translocations and multiplex ligation dependent probe amplification (MLPA) for the investigation of all telomere regions were the methods of choice. However, although introduction of FISH and MLPA improved the resolution of karyotyping in a locus-specific way, genome wide screening for deletions and duplications is not possible with high resolution using these techniques. Genomic microarrays provided the technology for screening the whole genome for copy number aberrations with a significantly improved resolution.

The genomic microarray-technology uses fragments of DNA sequences with a known genomic location, also called reporters or probes, which are fixed to a surface, the array. The first arrays were compiled of hundreds or thousands of Bacterial Artificial Chromosome (BAC) clones, harbouring a unique human genomic DNA sequence insert for a specific location on the genome as probes. On this chip, the target patient DNA labelled with a green fluorescent label and reference DNA with a red fluorescent label were then hybridised. Equal amounts of these labelled samples were mixed and hybridized to the array. During hybridisation, the patient and reference DNA will compete to bind to the complementary array probes. In the case of a deletion (loss) in the patient, less green-labelled DNA will bind to the array probe, showing a decreased green intensity for the specific probe(s). In the case of a duplication (gain), more of the green labelled DNA will bind, resulting in an increased green signal intensity. These signal intensity ratios are plotted by analysis software against the known position of the array probe in the

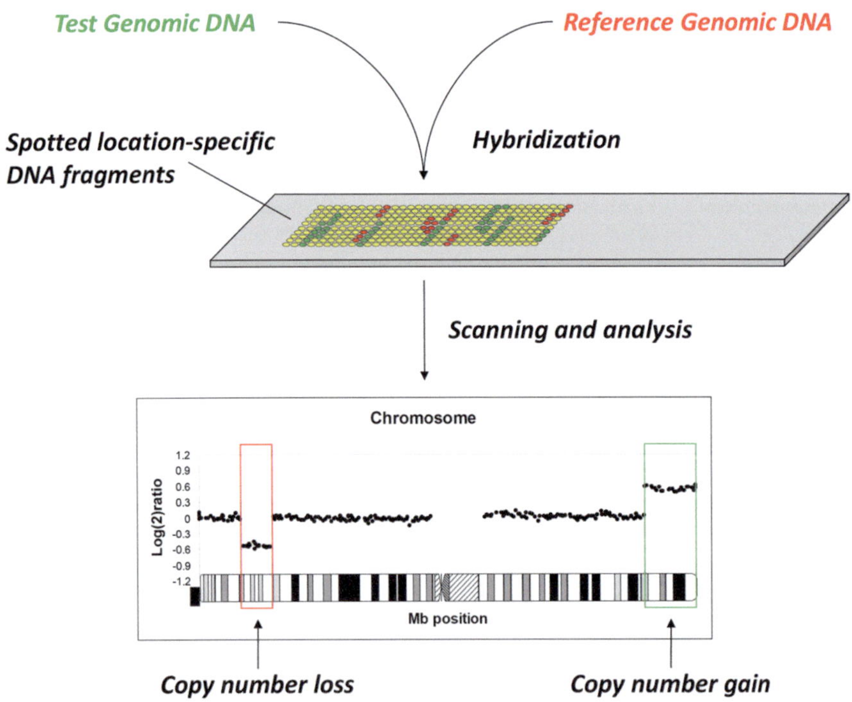

Fig. 8.1 Schematic representation of array comparative genomic hybridization. In case of comparative hybridization, a DNA sample is co-hybridized with a reference sample, both labelled in different colours, onto an array with spotted location-specific DNA fragments. Differences in normalized intensity ratios show loss of genomic regions (in this example less *green* signal compared to the reference sample, thus a more *red* coloured spot, as shown by the *red* rectangle in the plot) or gain of genomic regions (more *green* signal compared to the reference sample, thus a more *green* coloured spot, as shown by the *green* rectangle in the plot). Alternatively, several (commercial) array platforms use in silico reference panels instead of co-hybridization

genome, thus creating a visual profile for the whole genome and the regions that are present in aberrant copy numbers (Fig. 8.1).

Genomic microarrays have evolved greatly since these first experiments, and commercially produced arrays are now used for copy number investigations, which most often use anything from several hundred, up to several thousand specific short oligomers as probes for the detection of genomic imbalances.

8.3.1 SNP Arrays

One specific type of array, called the SNP array, uses oligomer probes which are selected around single nucleotide polymorphisms (SNPs). In this way it is possible to obtain genotype information about these SNPs in addition to copy number information. This genotype information can then be used to identify large regions of copy-number neutral homozygosity (ROH) (Fig. 8.2). When parents of the patient are consanguineous, several of these large ROH can be observed in the array profile. These regions can subsequently point to the presence of two recessive alleles, thus contributing to the elucidation of the genetic trait that causes the phenotype of the patient.

Moreover, the presence of one or a few large ROH on only a single chromosome in the genome may suggest a uniparental disomy (UPD). If UPD happens in a chromosomal region in which imprinting is a known mechanism for disease, such as the Prader Willi/Angelman syndrome region of chromosome 15, the finding of ROH could be a relevant lead to the presence of UPD and additional appropriate DNA diagnostics is then needed to confirm the diagnosis.

Comparable to the usefulness of ROH blocks in revealing the genomic location of a recessive trait in consanguineous couples, SNP arrays can also be instrumental in determining identity by descent of regions of interest in non-consanguineous families with multiple siblings with similar phenotypes, suggesting recessive disease. The genomic regions of which both alleles in two affected siblings have the same haplotype based on SNP content will most probably be the regions that will point to the recessive trait. The power of this type of screening will increase rapidly if one or two, affected or unaffected, siblings can be added to this comparison.

Over the last 10 years, genomic microarrays have replaced karyotyping as the first test in patients with ID and/or MCA or US anomalies, and diagnostics in certain types of haematological malignancies is currently shifting towards genomic microarrays as well. This technology can detect unbalanced chromosome aberrations with high resolution. However, for patients where you expect balanced genetic alterations (like a balanced translocation between chromosomes 13 and 14), classical karyotyping is still needed at this moment (Carter 2007).

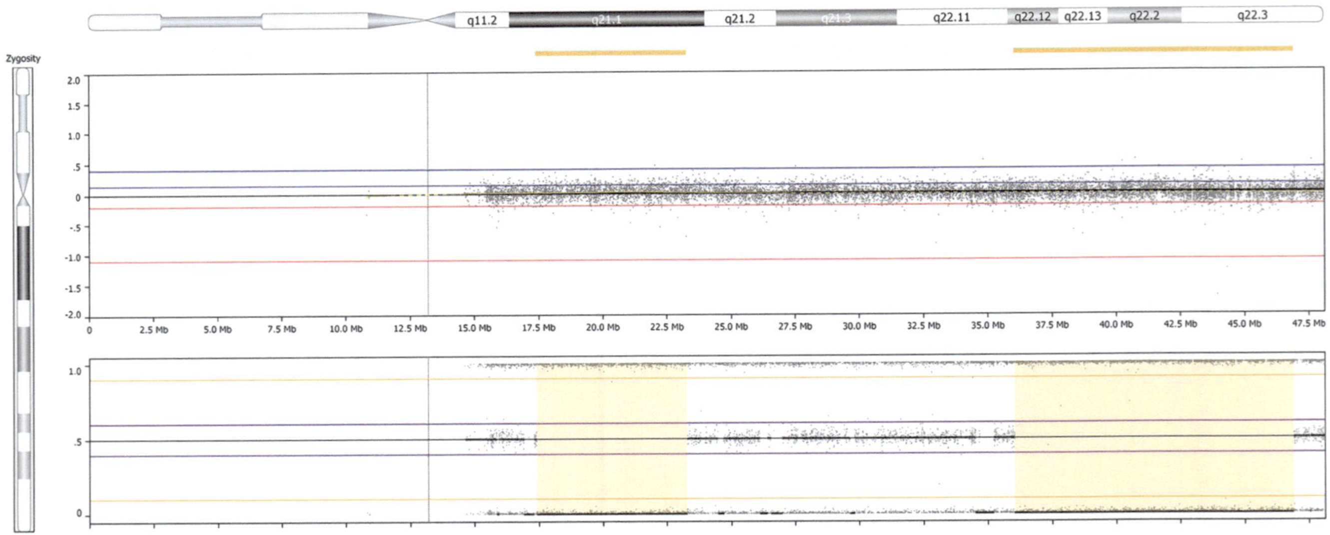

Fig. 8.2 SNP array result of the whole chromosome 21, showing a normal log ratio for the whole chromosome in the *upper panel* and two large regions of homozygosity (ROH) for the used SNP probes, depicted with the *yellow* blocks, in the *lower panel*. These ROH lack heterozygous AB probes, which are usually plotted around a 0.5 value and show only probes around the 0.0 value line (AA probes) or around the 1.0 value (BB probes)

8.3.2 *Example of a SNP Array in a Patient with Congenital Anomalies*

A six month old female patient was investigated using a microarray because of a congenital heart defect, hypotonia and dysmorphic features. The SNP array demonstrated the presence of a causal 3.5 Mb loss in the short arm of chromosome 17 (17p11.2) (Fig. 8.3). The deleted genomic region contains at least 100 annotated genes, including the *RAI1* gene. Deletions of this specific genomic region, or mutations in the *RAI1* gene, result in the Smith-Magenis syndrome, explaining the phenotype of the patient. Subsequent testing in the parents showed that the deletion arose de novo, which can be explained by a recombination between highly homologous repeat clusters in the genome directly flanking the loss. Also, the fact that the deletion was de novo, confirms that it is causative for the phenotype of the patient.

8.3.3 *Multiple Myeloma*

Multiple myeloma (MM), or Kahler's disease, is a malignant disease characterised by proliferation of clonal plasma cells in the bone marrow, where they interfere with the production of normal blood cells. The secretion of monoclonal immunoglobulins, which are detectable in the serum or urine, typically accompanies this proliferation of plasma cells.

In MM, typical genetic abnormalities are, amongst others, translocations involving the IGH gene (14q32) and copy number aberrations involving chromosomes 1, 9, 11, 13, and 17 (specifically the *TP53* gene locus). These aberrations are associated with prognosis and help decide the best therapy for patients. Until recently, all these targets were investigated using FISH. However, as analysing the FISH results for MM can be very complex and time consuming, more and more laboratories are now starting to use genomic arrays for detection of the relevant copy number aberrations (Fig. 8.4), whereas the IGH translocations are still being addressed using FISH.

8.4 Fragile X Syndrome

Fragile X syndrome (FXS) is named after the observation that a 'fragile site' on the long arm of the X-chromosome (Xq27.3) was observed in cultured cells of male patients. These males were mentally retarded and showed clinical symptoms like an elongated face, large protruding ears, large testes, and a variety of behavioural and social problems. Diagnostic cytogenetic testing was recommended for mentally retarded males since the observation of the Fragile X-chromosome. Unfortunately,

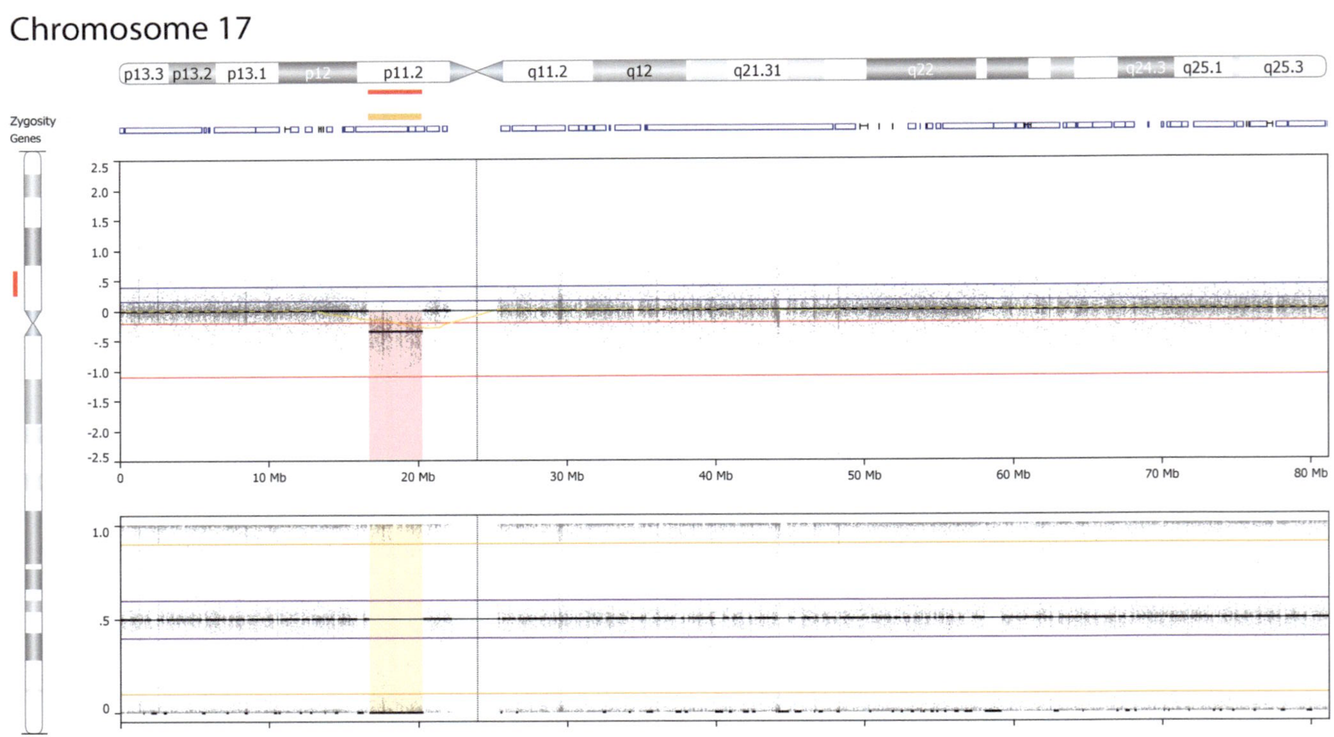

Fig. 8.3 Array result of chromosome 17. The *upper panel* shows the log ratio of individual SNP probes, plotted on their unique location on the chromosome. A probe average between the horizontal *red* lines will be called as a loss, whereas a probe average between the horizontal *blue* lines will be called as a gain. The *lower panel* shows the allele frequency of the tested SNP probes. In case of heterozygosity of the SNP, the frequency is plotted between the *purple* lines. The array result shows a 3.5 Mb loss in the short arm, depicted by the *red* block. Since there is only one copy left for this region, the allele frequency plot shows a region of homozygosity, depicted by the *yellow* block

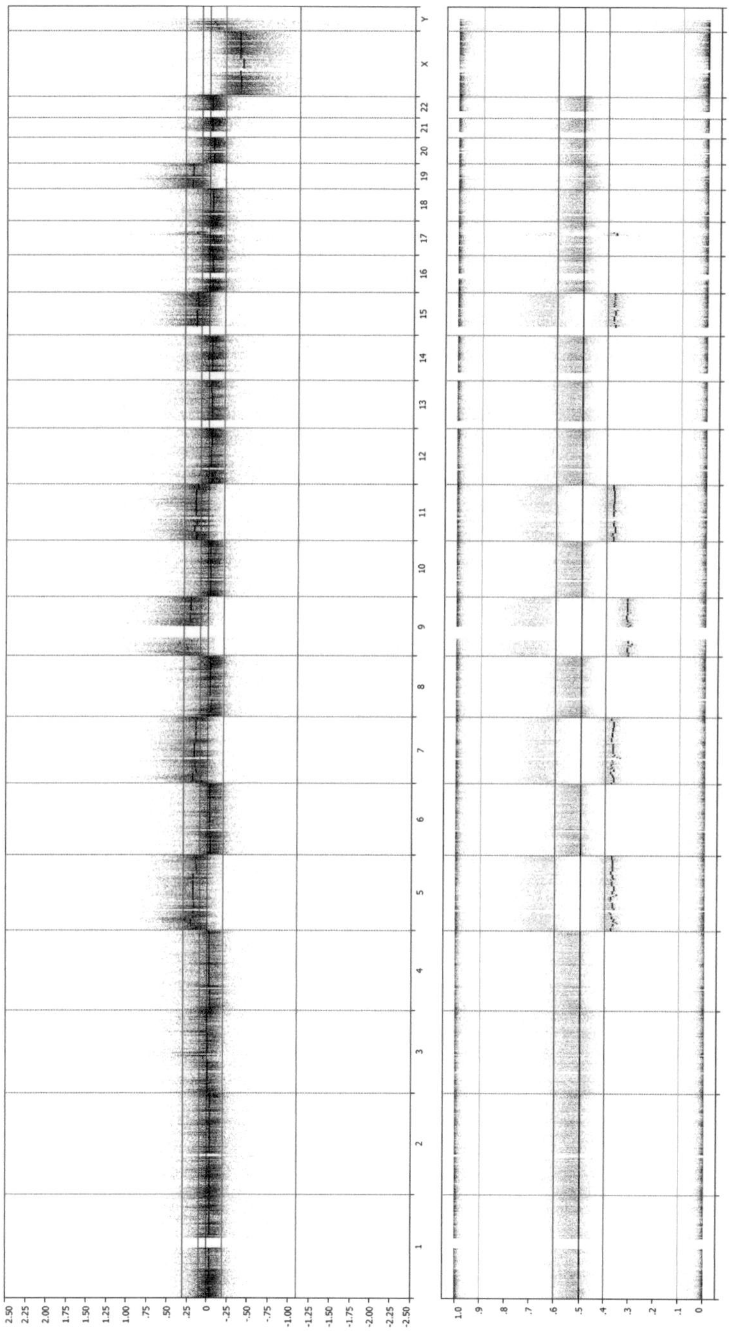

Fig. 8.4 A representative result of the whole genome of an array investigation for copy number alterations in MM. The different chromosomes are plotted along the x-axis. The result shows gains of the whole chromosomes 5, 7, 11 and 15 (3 copies) and 9 and 19 (4 copies)

the cytogenetic techniques were not reliable enough to determine carriers (females, FXS is an X-linked disorder) of this disease, limiting the analysis only to diagnosing male patients (both postnatal and prenatal).

In 1991, a CGG repeat in the 5′ UnTranslated Region (UTR) of the Fragile X Mental Retardation 1 (*FMR1*) gene was found to be the underlying cause of FXS (Verkerk et al. 1991). In the normal population, this repeat varies between 6 and 44 repeats (29 and 30 repeats are the most common). In patients, however, the CGG repeat was expanded to >200 repeats (called a full mutation). In addition, the number of triplets differs in each cell for this CGG repeat, a so-called somatic mosaicism. Further research testing showed that patients having a full mutation lack *FMR1* mRNA and therefore also lack the Fragile X Mental Retardation Protein (FMRP). The lack of transcription of *FMR1* is caused by methylation of the CpG island in the promoter of *FMR1*. This methylation pattern was not seen for *FMR1* with repeats in the normal, or premutation (between normal and full) range. This observation was of great importance, because the methylation appeared to be specific and therefore useful as target for diagnostic tests. It also made it possible to identify carriers. Carriers either have a premutation (55–200 repeats) or a full mutation (>200 repeats), with 45–54 repeats being an unclear, intermediate zone (grey zone). Around 50% of females with full mutations are affected with FXS (Kaufmann and Reiss 1999).

The gold standard to determine the number of CGG repeats for *FMR1* is Southern blotting. Southern blotting is a technique where genomic DNA is digested with specific restriction enzymes (e.g. HindIII in combination with EagI) and hybridized to a gene-specific probe (e.g. pP2). This technique has the advantage that it can detect both an expanded repeat and the methylation status of the repeat in a single experiment. The enzyme HindIII is used to obtain smaller fragments. The restriction site of EagI is methylated in FXS patients and is therefore not recognized and used by EagI. This results in larger fragments compared to normal individuals in the digestion profile of FXS patients.

In Fig. 8.5 the results are shown of a Southern blot experiment in a family with FXS patients. Digestion using the enzymes HindIII/EagI will result in a fragment of ~2.8 kb (normal (N), unmethylated) for normal males because the EagI restriction site is unmethylated. For normal females a ~5.2 kb band (normal (N), methylated) will be visible in addition to the ~2.8 kb band, because in females one of the two X-chromosomes is inactive (i.e. methylated) in each cell. In FXS males a smear (full mutation (F)) is observed on the blot, which is the result of the presence of a large repeat that differs in size between cells and the fact that the EagI restriction site is methylated. Female FXS patients show a band for the normal X-chromosome and a band for the pathogenic expanded CGG repeat (full mutation).

Because FXS is an X-linked disorder, a mother of an FXS patient will always be a carrier. The full mutation is clearly visible for both son and daughter. The length of the CGG repeats in the mother is shorter (premutation) compared to that of her children (full mutation). This example illustrates the instability of the *FMR1* CGG repeat. In most cases, the repeat size of the mother is smaller compared to her offspring, however, in some families the mother herself has a full mutation. Also in

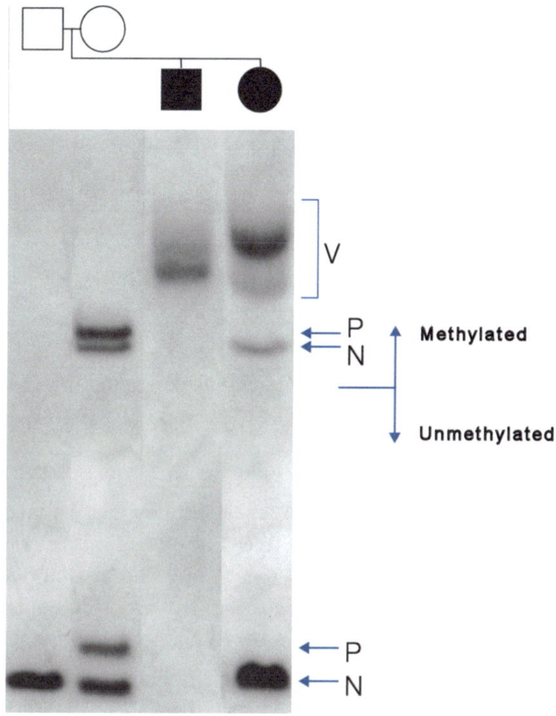

Fig. 8.5 Southern blot analysis of an FXS family. Genomic DNA is digested using the restriction enzymes HindIII/EagI and hybridized using the probe pP2 specific for *FMR1*. *open square* = male, *open circle* = female; solid symbols = FXS patient; F = full mutation; P = premutation; N = normal

these families the size of the CGG repeat gets smaller as individuals higher in the pedigree are tested.

The exact length of the *FMR1* CGG repeat cannot be determined using Southern blot. Instead, PCR analysis is used. Amplification is performed using deazanucleotides, because of the presence of many C and G nucleotides (see Vol. 1, Chap. 5). This conventional PCR test is not suitable to amplify fragments with >150 CGG repeats. Therefore, this test was not used for diagnostic testing of FXS patients but only for testing family members. The conventional PCR test was fortunately followed by improved PCR based tests and MLPA based tests that were able to detect the full mutations as well. Nowadays commercial kits using such methods are available. In Fig. 8.6 an example is shown of a female having a normal allele with 25 repeats and a premutation of 95 repeats. One important reason why the labour-intensive Southern blotting technique was, at least in many laboratories, not yet replaced with the fast and cheap PCR/MLPA based tests, was the fact that the methylation information was missing in these tests compared to Southern blotting. This limitation was solved and now (commercial) methylation sensitive PCR and MLPA test are available.

Figure 8.7 shows an example of a sample (process control sample, not a real individual) with several normal and premutation peaks and one full mutation peak (upper panel). The methylation data is indicated in de lower panel, showing that the

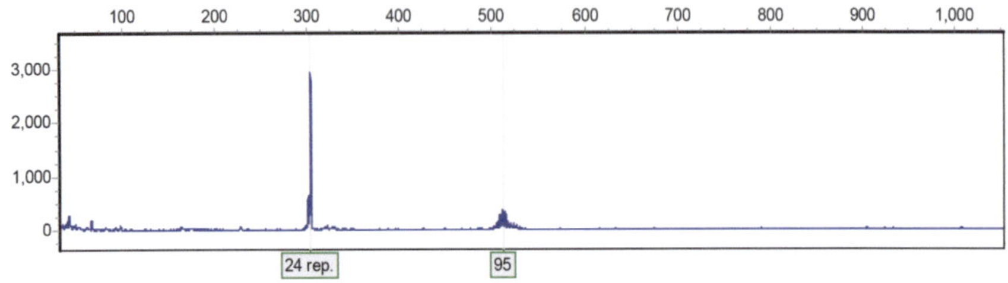

Fig. 8.6 Commercial PCR based CGG test showing a female having a normal allele with 25 repeats and an allele with a premutation of 95 repeats

18, 32, 86 and 225 CGG repeats are unmethylated, the 30 and 56 repeats are partially methylated, and the 116 repeats are fully methylated. Most laboratories now no longer use the Southern blotting technique and moved to a two-step approach in which first a PCR based analysis is performed to detect normal, premutation and full mutation repeats, followed by a methylation sensitive PCR test for those individuals having either a premutation or a full mutation.

Expansion of the CGG repeat and methylation of the promotor region of *FMR1* is observed in most FXS patients. In a very small number of FXS patients, however, a (intragenic) deletion of *FMR1* or a point mutation in *FMR1* is found.

Additional information about diagnostic FXS testing and Fragile X-associated disorders can be found in the official Fragile X syndrome best practice guidelines, which were approved by the European Molecular Genetics Quality Network (EMQN).

8.5 Hereditary Breast/Ovarian Cancer Syndrome (HBOC)

In the Netherlands, 1 out of 8 women will develop breast cancer, of which 5–10% is in a familial setting. Two genes were identified, *BRCA1* and *BRCA2*. When there is a mutation in one copy in one of those genes, this results in a high risk for the female carriers to develop breast and/or ovarian cancer. *BRCA1*, containing 23 exons (RNA is 7.8 kb), encodes a protein that is involved in DNA replication and in repair of double strand breaks, which are the result of for example ionizing radiation or exposure to certain chemicals. *BRCA2* is also a large gene. It contains 27 exons, coding for an mRNA of 10–12 kb. The protein product of *BRCA2* is involved in the same pathway as the *BRCA1* protein (Palma et al. 2006).

The inheritance mode for these two genes is autosomal dominant, meaning that offspring of a carrier have a 50% chance of inheriting the mutation. Since inactivation of both alleles is observed in tumours, it is clear that these genes are tumour suppressor genes.

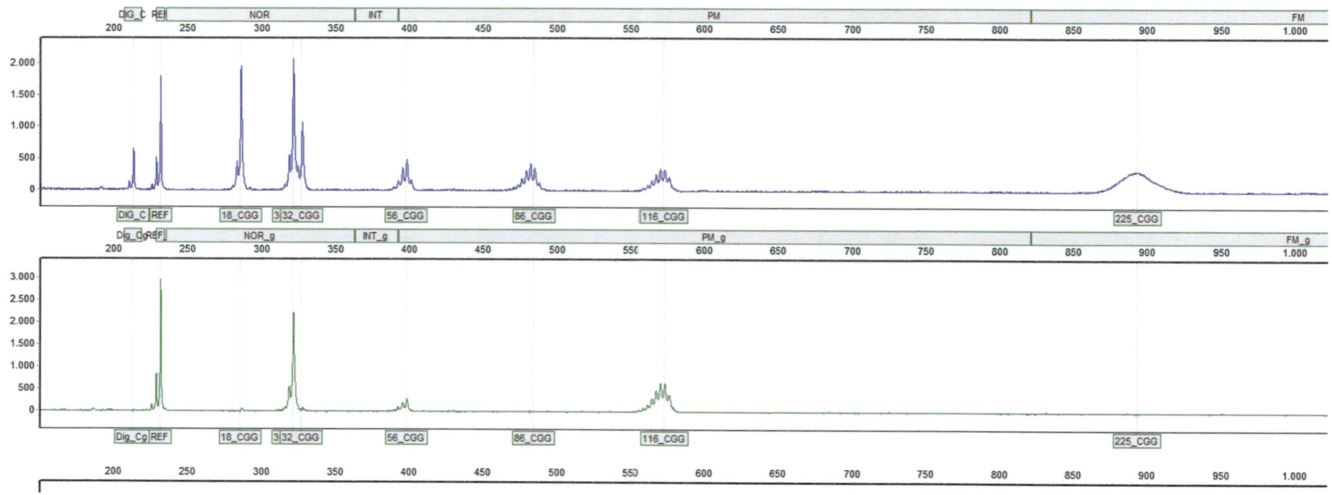

Fig. 8.7 Commercial methylation sensitive PCR based CGG test showing (*upper panel*) a process control sample with normal peaks (18, 30 and 32 repeats), premutation peaks (55, 86 and 116 repeats) and one full mutation peak (225 repeats). *Lower panel*: methylation status; unmethylated: 18, 32, 86 and 225 repeats; partially methylated: 30 and 56 repeats; fully methylated: 116 repeats

The chance to identify a pathogenic mutation depends on several clinical and familial characteristics, including the number of (young) breast cancer patients in a family and the tumour type: patients with a triple negative tumour have a higher chance of being a carrier of a pathogenic mutation, especially a *BRCA1* mutation. After performing mutation analysis of these genes in a lot of families, it is clear that there is hardly any clinical difference between a *BRCA1* and a *BRCA2* family. If the pathogenic mutation in a family is identified, predictive testing is offered to family members. If a healthy female is not a carrier of the familial mutation, the risk of this person is reduced to almost the population risk of developing breast/ovarian cancer, in the Netherlands 1:8. However, if she is a carrier, she has a lifetime risk of developing breast cancer of about 60–80% and for ovarian cancer it will be about 40%. These females can opt either for mastectomy and ovariectomy or regular medical surveillance.

Pathogenic mutations in *BRCA1* and *BRCA2* can either be small point mutations, identified by e.g. Sanger sequence analysis, or large rearrangements. Large rearrangements are in most cases identified by using *MLPA* (Multiplex Ligation-dependent Probe Amplification; see Vol. 1, Chap. 5). In *BRCA1* and *BRCA2,* all types of mutations can be identified—nonsense, frameshifts, missense, splice site and even silent nucleotide changes can be pathogenic. A nonsense mutation is the effect of a nucleotide change, which results in a premature stop codon with the result that either the RNA is not stable or a very short non-functional protein is produced. A frameshift mutation is a deletion or duplication of more nucleotides as long as it is not a number of nucleotides which can be divided by 3 (as that would not result in a frameshift). In most frameshift mutations, a premature stop codon will be introduced. A missense mutation is the result of one amino acid, which is replaced by another due to the nucleotide change. Sometimes a nucleotide change will not lead to a replacement of an amino acid, but it could still be a pathogenic mutation because it could have an effect on RNA splicing.

To identify a sequence change in a gene, primers are designed which are specific for the region that needs to be amplified. In case of *BRCA1* and *BRCA2* there are a few exons that are too large to be amplified in 1 amplicon. In these cases, overlapping PCR products need to be analysed. Most laboratories have an automated sequencer and the data can be transported directly into analysis programs. All these software programs use a comparison method that alligns the test result to a standardized reference sequence. The software program lists each difference between the reference and the test sequence. Common variants that do not result in clinical symptoms, so called neutral variants, are present in SNP databases. However, it is a challenge to interpret variants, which are not present in these SNP databases. To share data, gene specific mutation databases are being developed [e.g. Leiden Open (source) Variation Database (LOVD)].

A nice example of how complex the interpretation can be is given for the sequence change c.5096G > A, p.Arg1699Gln, at nucleotide position 5096 of the coding region of *BRCA1,* Guanine was replaced by Adenine, resulting in a replacement of the amino-acid Arg (Arginine) by Gln (Glutamine) (Fig. 8.8). This missense change was identified in several families of Dutch origin, but it did not

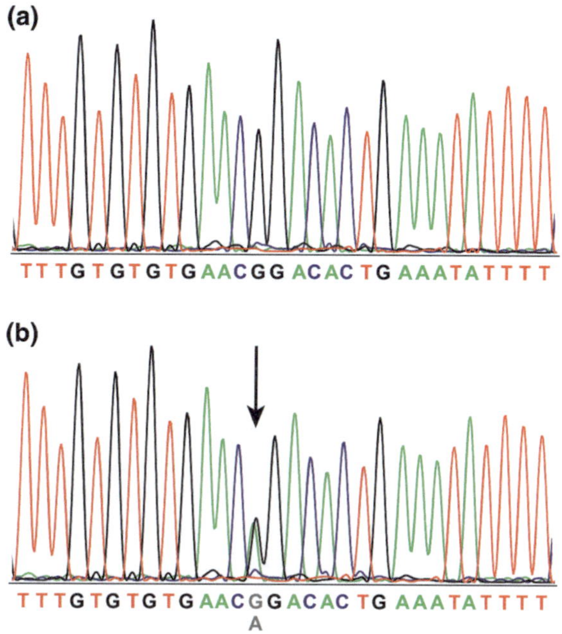

Fig. 8.8 Sequence result (electropherogram) of a part of exon 18 of *BRCA1* of a normal DNA (**a**) and DNA of a patient heterozygous for the c.5096G > A, p.Arg1699Gln pathogenic mutation (**b**). Under each peak, the dinucleotide detected (A, C, G or T) is given. The *arrow* above the sequence of the patient is nucleotide position c.5096. At this position, the DNA of the patient is heterozygous for the A and G dinucleotide. This nucleotide changes result in the replacement of the amino-acid Arg (Arginine) by Gln (Glutamine) at protein position 1699

completely segregate with the phenotypes in the families. An international collaboration was necessary to obtain enough families and tumour material. Functional tests were also performed and it turned out that this sequence change was a pathogenic mutation with reduced penetrance. This means that a lower number of female carriers will develop breast/ovarian cancer compared to females who are carrier of for example a pathogenic nonsense mutation in *BRCA1* (Spurdle et al. 2012). It is of course very important to be sure that the sequence change that is identified is classified correctly, because women are making choices based on these results.

8.6 Whole Exome Sequencing for Heterogeneous Disorders

The technique of massive parallel sequencing (next generation sequencing) is explained in Vol. 1, Chap. 8. Whole Exome Sequencing (WES) is a form of massive parallel sequencing where, prior to sequencing, all coding regions of the DNA (the exome, all exons) are enriched. The enrichment can be done by hybridization of fragmented genomic DNA to a chip, where all exons are represented by oligomeric probes, followed by washing steps to remove non-bound DNA and elution of the bound DNA. Alternatively, in-liquid hybridization to probes bound to small biotin-labelled spheres allows the extraction of the exome sequences using streptavidin antibodies. The enriched DNA is then used as input for massive parallel sequencing. After sequencing, analysis of the raw data starts (see Vol. 1, Chap. 9 on bioinformatics for detailed information). The so-called "reads" are mapped back to the human genome. This generates a BAM file (a binary format to store sequence data), which can be analysed in different ways. One can now identify the changes between the exome that has been sequenced and the reference genome, both for single nucleotide changes as well as for (small) deletions and/or insertions. The output of this analysis is called a vcf file (variant calling file). Because the development of the analysis tools has been going very rapidly, and has also been running in parallel in several institutions, the format of vcf files can differ, depending on the platform they have been generated on. An important parameter is coverage: the number of reads that are mapped to the same position in the genome. Usually, a minimal coverage of 10x (meaning that there are at least 10 reads for each position in the genome) is taken as the lower limit for a reliable detection of SNPs. For a heterozygous SNP, this means that, on average, five reads will have the reference nucleotide in that position, and five reads will have the alternative nucleotide, but of course there is always a chance (albeit smaller) that this is 3 and 7, or even 1 and 9.

A minimal coverage of 10x means that the average coverage should be 50x or even higher for diagnostic purposes, because there will always be difficult exons (i.e. difficult to sequence for one reason or the other) that are underrepresented. On the other hand, there are exons that are always covered very well. One way to tackle this problem is by improving the arrays that are used to select the exome sequences at the beginning of the process, and this has solved part of it. The final solution is of course to skip the selection and sequence the whole genome. This can then be followed by bioinformatic selection of the exome and the variants in the exome. It is foreseen that this will happen in the near future. At the moment, sequencing costs and the challenge of handling such big data sets preclude this option.

After generating the vcf file, the filtering process and variant interpretation can start. Several tools can be used for the interpretation of variants. In academic institutions, the pipeline is sometimes home build, but also commercial pipelines are available, like Cartagenia. Basically, the interpretation of the variants starts with annotation. For each variant it is established whether it has been seen before in one of the online databases, whether the variant has a population frequency (i.e. how many people in the general population also have this variant), if there is a consequence of the variant on the coding sequence (i.e. does it lead to an amino acid change, or a stop codon). Based on these annotations, the interesting variants can be selected. For instance, one can choose to look at only the variants that are present in the genes that are associated with the disease of the patient that is investigated. Furthermore, all the variants that have a frequency of >1% in the population can be skipped, because those are very likely to be neutral polymorphisms. Using these steps usually reduces the number of variants that have to be looked at to less than 20 (Fig. 8.9).

For patients with intellectual disability, the approach is a bit different. Because it has been shown that in a large proportion of cases the causative mutation is de

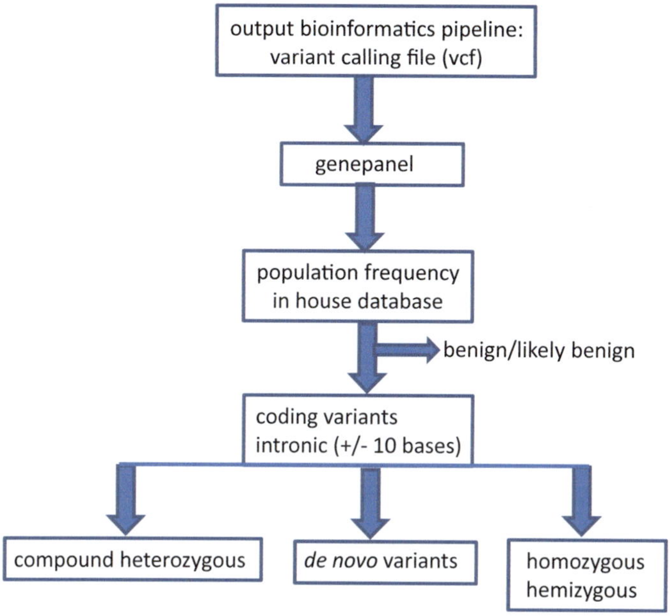

Fig. 8.9 Common filtering strategy used in the laboratory using either a homemade bioinformatics filtering pipeline or a commercial software program (e.g. Cartagenia)

novo, the first line approach is to compare the vcf file of the patients with that of the parents. A filter is applied for those variants that are present in the child, but not in the parents. In typical cases, one ends up with 1–3 de novo variants, from which it is easy to determine which variant should be the causative one. When no de novo mutations are found, filtering for two variants in the same gene can be applied. These variants need to be compound heterozygous, i.e. both present in the child, where one parent is carrier of one mutation, and the other parent is carrier of the second mutation. Thirdly, homozygous and hemizygous variants (when dealing with a male child) should be checked. Hemizygous mutations on the X-chromosome can only be inherited from the mother. In case of a homozygous mutation, both parents should be heterozygous for the sequence change.

The enrichment for exome sequences has evident advantages. Because only 1% of the human DNA is coding DNA, much less sequence has to be generated to sufficiently cover all regions of interest, which is much more costs effective. Furthermore, bioinformatics analysis of the amount of data representing only 1% of the genome is much easier than analysing the files that are generated with whole genome sequencing (WGS). Lastly, mutations found in the coding regions explain the majority of cases. It is expected that mutations in non-coding regions are far less likely to be causative.

In general, it is good practice to inform patients of the risks of genetic tests beforehand. This has become even more obvious with the introduction of WES.

One of the major issues with this kind of approach is the risk of unsought findings. The classic example is finding a pathogenic mutation in one of the breast cancer genes in a patient that has a question about something completely different, for instance a heart condition. There are several ways to deal with this: first of all, to make sure that these mutations do not show up in the analysis, by selecting only those genes associated with (in this case) the heart condition. In cases where this cannot be avoided, it is very important that patients are aware of the risks of finding something that is not expected, and that can have a major impact on their lives (Stranneheim and Wedell 2016).

8.7 Conclusion

In this chapter the field of Clinical Genetics is discussed. The field has expanded tremendously in the last 30 years due to the ever more rapid discovery of the genes involved in genetic disorders. In the past 15 years, the diagnostics of hereditary breast/ovarium cancer has taken up much of the resources of the genetic laboratories. CNV detection has been the method of choice for patients with intellectual disabilities. With the arrival of massive parallel sequencing in the diagnostic setting, the focus is shifting to "one test fits all". Patient awareness of the possible outcomes of these tests is paramount before starting the investigation.

Further reading

Carter NP. Methods and strategies for analyzing copy number variation using DNA microarrays. Nat Genet. 2007;39(7 Suppl):S16–21.

Kaufmann WE, Reiss AL. Molecular and cellular genetics of fragile X syndrome. Am J Med Genet. 1999;88(1):11–24.

Palma M, et al. BRCA1 and BRCA2: the genetic testing and the current management options for mutation carriers. Crit Rev Oncol Hematol. 2006;57(1):1–23.

Spurdle AB, et al. BRCA1 R1699Q variant displaying ambiguous functional abrogation confers intermediate breast and ovarian cancer risk. J Med Genet. 2012;49(8):525–32.

Stranneheim H, Wedell A. Exome and genome sequencing: a revolution for the discovery and diagnosis of monogenic disorders. J Intern Med. 2016;279(1):3–15.

Turnpenny PD, Ellard S. Emery's essential elements of genetics, 14th edn. London: Elsevier Churchill Livingstone; 2011.

Verkerk AJ, et al. Identification of a gene (FMR-1) containing a CGG repeat coincident with a breakpoint cluster region exhibiting length variation in fragile X syndrome. Cell. 1991;65 (5):905–14.

Chapter 9
Molecular Diagnostics in Plant Health

Peter Bonants and René te Witt

Abstract In several crops, their quality and condition is very important. The products are relatively expensive and plant diseases can destroy valuable harvests in a very short time. The presence or absence of plant pathogens and pests, therefore, needs to be accurately tested. For decades, identification and detection methods have been designed to demonstrate that starting material for further propagation and also certified bulbs, tubers, cuttings, and plants are free of many plant pathogens. Every year, millions of plants are tested, including approximately one and a half million bulbs. Despite the ease of low-cost, fast, and well-automated test methods, such as ELISA, more and more molecular-based techniques are used the last decade. These techniques are, generally, more specific and more sensitive. These new tests are also necessary because plant pathogens are often latent (asymptomatic) present in the material to be tested. The Netherlands has a high in—and export of plants and plant products. These large trading volumes may harbour many plant pathogens. The European Commission has drawn up a list of unwanted plant pathogens (quarantine organisms). This list contains more than 300 different organisms (viruses, bacteria, fungi, phytoplasmas, nematodes and insects). Each EU-country has to control the in—and export traffic of plants and plant products for the introduction of these unwanted organisms. In the Netherlands, this task is performed by four inspection services: Naktuinbouw (Netherlands Inspection Service for Horticulture), NAK (The Dutch General Inspection Service), BKD (Inspection Service for Flowerbulbs) and KCB (Quality Control Bureau). The NVWA (Dutch Food and Consumers Authority) is the NPPO (National plant protection organisation) in the Netherlands and supervises the inspection services. Many different molecular methods are used daily for the detection and identification of many plant pathogens and plant diseases.

P. Bonants
Business Unit Biointeractions & Plant Health, Wageningen UR, Wageningen, The Netherlands

R. te Witt (✉)
NMDL, Rijswijk, The Netherlands
e-mail: rene.te.witt@nmdl.nl

E. van Pelt-Verkuil et al. (eds.), *Molecular Diagnostics*,
DOI 10.1007/978-981-10-4511-0_9

Keywords Molecular diagnostics · Quarantine · Plant pathogens · Detection · Identification

9.1 Introduction

In this chapter, a number of PCR-based applications that are used in the field of Plant Health during cultivation and testing of plants and plant products will be discussed. Many different plants and plant products (flowers, flower bulbs, seed potatoes, trees, seedlings, vegetables, fruits and many other) can carry different pathogens (viruses, bacteria, fungi, nematodes, insects and phytoplasmas). Each of these pathogens has its own characteristics and needs to be detected in an early stage. A detection method can detect a single target (singleplex) or several different targets (pathogens) simultaneously (multiplex). In these tests, diagnostic specificity and diagnostic sensitivity are important (see Vol. 1, Chap. 6). Detection can be quantitative (the load of a pathogen in a sample) or qualitative (presence/absence). Also tests can be designed to discriminate between living or dead pathogens or between different genera or pathotypes or *formae speciales* of a certain pathogen. In addition, many different methods have been designed to identify a pathogen to species or subspecies level. This is important since the EU quarantine list is species or subspecies specific (Table 9.1).

Table 9.1 List of quarantine species and their status, which were used in the EU project Portcheck

List	Quarantine organism	Common name	EU status
A	*Phytophthora ramorum*	Sudden oak death (A1/A2 (US vs EU mating types)	–
	Bursaphelenchus xylophilus	Pinewood nematode	IIAI
B	*Ralstonia solanacearum*	Potato brown rot	IAII
	Clavibacter michiganensis subspp. Sepedonicus	Potato ring rot	IAII
	Tilletia indica	Karnal bunt	IAI
	Potato spindle tuber viroid	PSTVD	IAI
	Thrips palmi	Palm Thrips	IAI
C	*Synchitrium endobioticum*	Wart disease	IAII
	Meloidogyne chitwoodi/M. fallax	Rootknot nematodes	IAII
	Guignardia citricarpa	Citrus black spot	IIA1
	Globodera pallida/rostochiensis	PCN	IAII
	Beet necrotic yellow vein virus	Rhizomania	IB
	Tomato spotted wilt virus	TSWV	IB
	Pepino mosaic virus	PepMV	–
	Colletotrichum acutatum	Strawberry blackspot	IIAII

This chapter describes examples of EU projects in which a wide variety of plant pathogens have been identified and/or detected in different substrates.

9.2 Detection: What, Where and How?

Quality monitoring is necessary in order to ensure safe and sustainable agriculture within Europe. In order to effectively implement quality monitoring at all levels of the production chain, European agriculture should develop harmonised methods, tools and products for detection, classification, control and inspection of plant pathogens or plant-related organisms, especially quarantine organisms (Q-organism). Identification and detection methods are applied to detect a large set of plant pathogens. These methods are of great importance in research and development towards the design of effective environmental friendly control measures based on resistance breeding, cultural practices, and/or biological control. In addition, identification and detection are essential for studying the epidemiology of pathogens, to control and to monitor quality and to obtain certification of crops. The question from the market for reliable identification and sensitive detection of plant pathogens and beneficial microorganisms increases. Important developments nowadays are:

- Quantitative detection to quantify inoculum densities of pathogenic variants in samples to estimate risks for economic damage and
- Multiplex detection for multiple plant pathogens (targets) in the same sample.

When validating a method, different aspects have to be considered and certain criteria have to be fulfilled to develop an effective detection method:

- Diagnostic specificity: The test should be very specific and only detect the target pathogen and not closely related species; no false positives should be observed.
- Diagnostic sensitivity: The test should be sensitive enough to detect e.g. 1 zoospore; no false negatives should be observed. This counts especially for Q-organisms, for which a zero-tolerance is obligatory.
- Costs: The test should be cost effective for routine testing.
- Expertise: The presence of sufficient technical and personal expertise needed to perform the test, should be considered.
- Robustness: The test should be robust, including a high repeatability and reproducibility.
- High throughput: The test should be suitable for high throughput screening.
- Speed: The results should be obtained within a convenient period.

Many detection methods have been developed for different kinds of substrates, e.g. soil (including potting compost), water (e.g. water for irrigation or in recirculating water systems), plant material, inoculum (tests for the purity in research; biological control agents) and for different kinds of organisms:

- Viruses and viroïds: viruses of vegetables, fruits, ornamentals, potatoes.
- Bacteria: brown rot (*Ralstonia*), ring rot (*Clavibacter*), root knot (*Agrobacterium*), soft rot (*Erwinia*)
- Phytoplasmas: *Stolbur*
- Fungi: *Phytophthora* spp., *Fusarium oxysporum*, *Rhizoctonia solani*, *Synchytrium*, *Olpidium*, *Phoma*, *Alternaria*, *Nectria*, *Guignardia*
- Nematodes: *Meloidogyne chitwoodi*, *M. fallax*, *M. hapla*, *M. incognita*, *M. javanica*, *M. arenaria*, *Globodera pallida*, *G. rostochiensis*
- Insects: *Thrips* species, *Bemisia tabaci* (silverleaf whitefly).

Both EU and national regulations describe a wide variety of recommended techniques for the detection of plant pathogens. Currently, each pathogen or group of pathogens present in a host plant requires its own specific assay, which can be based on:

- Physiological and biochemical features (e.g. Biolog: growth on different nutrients etc.)
- Biological features on bait plants (e.g. virulence tests on different cultivars)
- Morphological features (e.g. microscopy: binocular, TEM, SEM) after plating on media.
- Protein/carbohydrate-level.
- Antisera (mono—and polyclonal antibodies) can be used:
  - In ELISA for detection purposes of viruses. For 5 potato viruses (PLRV, PVY, PVX, PVS and PVA) millions of tests are performed each year in the Netherlands to screen seed potatoes for certification and export.
  - In immunofluorescence-colony staining (IFC) for the detection of certain bacteria. After incubation of the bacteria on agar fluorescent-labelled antibodies are added. After washing unbound antibodies fluorescent-labelled colonies can be visually detected under an UV microscope and isolated for further studies, e.g. for confirmation by PCR.
- Isozyme-patterns e.g. malate dehydrogenase (MDH), malic enzyme (MDHP), esterase, glucose-6-phosphate isomerase and peptidase.

Since the introduction of amplification methods for nucleic acids and the increasing availability of sequence data, several methods have been developed for detection and identification purposes:

- On RNA level: RT—qPCR, NASBA or LAMP using primers and probes (Molecular Beacons, hydrolysis probes)
- On DNA level: PCR (qPCR, ITS-RFLP, SCAR-PCR, AFLP, multiplex PCR, (nested) PCR, RAPD-PCR) using primers and probes (Molecular Beacons, hydrolysis probes), REP-PCR, ISSR, SSR, (RT)-PCR, DGGE, NASBA, LAMP).

The amplicons can be detected by traditional agarose gel electrophoresis using an intercalating dye (e.g. ethidium bromide) or in "real time" using fluorescent probes. For instance the hydrolysis probe, which consists of two fluorescent groups: a

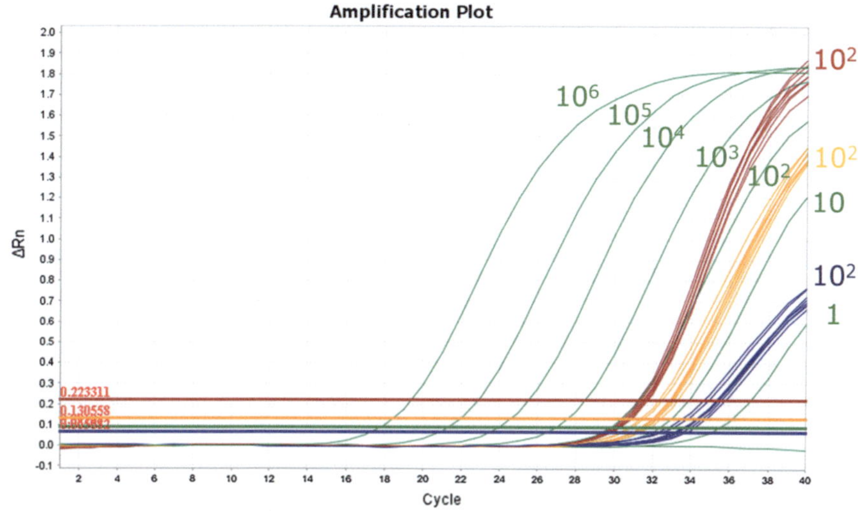

Fig. 9.1 Multiplex qPCR amplification curves for simultaneous detection of four targets. The probes for the four targets are labelled with different fluorophores and are shown in different colours in the figure. Copy numbers for the four targets are different and are indicated

reporter and a quencher (see Vol. 1, Chap. 3). An example of multiplex analysis using hydrolysis probes is shown in Fig. 9.1. The different targets can be detected simultaneously in one amplification reaction by using a maximum of four hydrolysis probes that are labelled with four different fluorochromes; three different targets and one to detect the internal control.

9.3 Detection Methods for *Phoma* Species

Stagonosporopsis andigena (formerly named *Phoma andigena*) and *Stagonosporopsis crystalliniformis* (formerly named *Phoma crystalliniformis*) are both quarantine fungi found in the Andes, the origin of the potato. Both can affect potato and tomato. For detection of these pathogens outside this area, a robust and rapid detection method is required.

Current detection methods are generally based on the Internal Transcribed Spacer (ITS) sequences. However, the variability of the sequence of this intergenic region is not high enough to obtain acceptable resolution between these species and to develop a specific PCR method. Therefore actin-encoding sequences were determined between various relevant *Phoma* species on potatoes (Andes region). After DNA isolation from 313 isolates, DNA amplification was performed on β-actin using consensus PCR primers described by Aveskamp et al. (2008). DNA sequencing was performed on the amplicons using standard Sanger sequencing methods. Alignment of the sequences resulted in a phylogenetic tree as shown in

Fig. 9.2. Based on those 313 B-actin sequences, primers and probes were developed for six individual *Phoma* species. Table 9.2 shows that the six developed qPCRs are specific for each species. As an example, Fig. 9.3 shows an amplification curve for *P. andigena*. These qPCRs have been tested on a large number of related *Phoma* species in order to determine the diagnostic specificity. After this, DNA extracts of potato/tomato leafs, which were spiked with mycelium of the two relevant species *S. andigena* and *S. crystalliniformis* were used to confirm the required specificity.

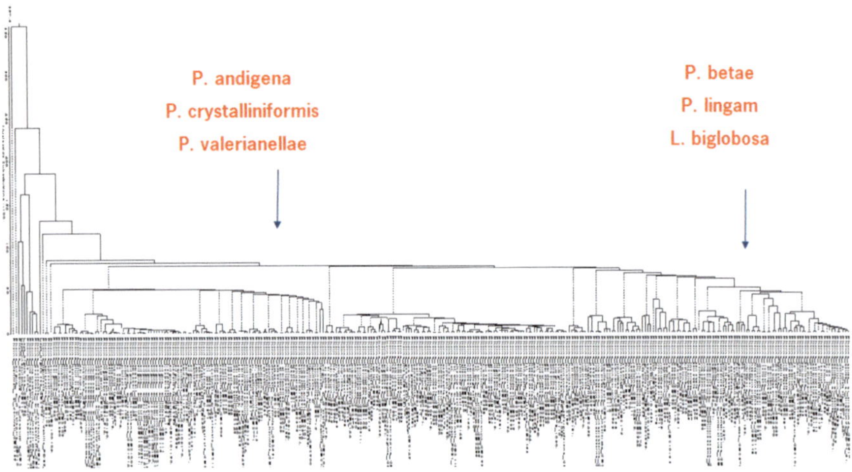

Fig. 9.2 Phylogenetic tree based on 313 B-actine sequences of Phoma isolates

Table 9.2 Specificity of qPCR's for six Phoma species. Cq values of the specific assays are indicated

Phoma	CBSnr	P.andigena F2xR1 +P1	P.crystalliniformis F2xR1 +P1	P.betae FxR +P1	P.valerianellae FxR +P1	P.lingam FxR1 +P1	L.biglibosa F1xR1 +P1	L.biglibosa F1xR3 +P3
andigena	269.80	19.38						
crystalliniformis	713.85		19.91					
betea	523.66			20.42				
valerianellae	273.92				19.63			
lingam	147.24					19.27		
biglobosa	475.81						23.06	24.4
subbolthauseri	104.42							
labilis	124.39							
briardii	101635							
terricola	165.89							
pimpinellae	101637						33.08	
intricans	139.78							
radicina	111.79							

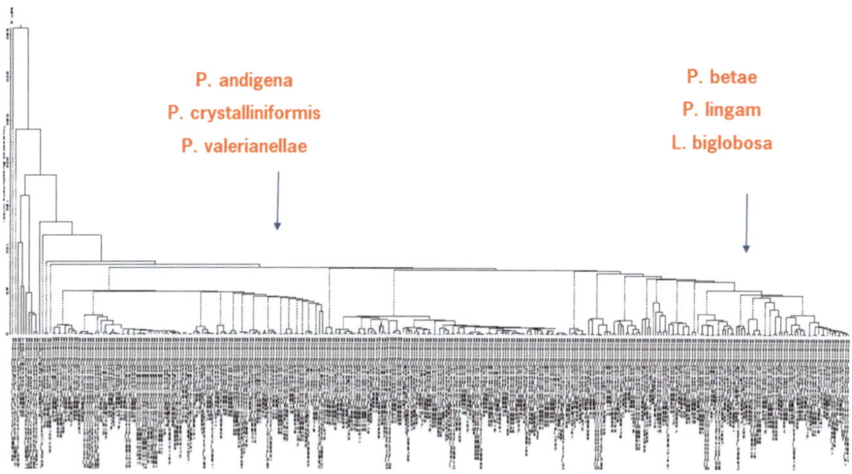

Fig. 9.3 Amplification curves of qPCR for 2 isolates of Phoma andigena, tested in duplicate

9.4 EU Project Portcheck

Portcheck was an FP6 EU project, which provided tools and procedures to EU member states that allow Plant Health competent laboratories and inspection services to perform molecular diagnostic assays "on-site" and at points of entry. The project developed and evaluated qPCR assays for a number of harmful key organisms, including the Oomycete *Phytophthora ramorum* (sudden oak death) and pinewood nematode. These assays were transferred to portable qPCR platforms, which were originally developed for bio-warfare and bio-terrorism applications. Particular attention was paid to the problems associated with sampling and nucleic acid extraction in field conditions. The project mobilised an impressive consortium, including both academic and SME (small and medium-sized enterprises), research partners, along with probably the most comprehensive and most representative network of official Plant Health laboratories and inspection organisations ever assembled for an EU RTD proposal. The consortium also included partners with a proven track record of developing and successfully deploying on-site diagnostic kits to official Plant Health inspection services. Consultation with the principal stakeholders of the technology (inspection services, competent laboratories, DG SANCO (Directorate General for Health and Food Safety of the EU), trade representatives etc.) throughout the project was of paramount importance. Adoption of the new technology represented a revolution in the way Plant Health services performed inspections in support of Council Directive 2000/29/EC and contributed to the reduced risk of the import (and export) and establishment of harmful organisms. This policy contributed to prevent any consequential negative impacts on the sustainability of European agricultural systems, to protect critically important natural forest ecosystems and other negative economic, social, and environmental effects associated with the establishment of non-indigenous plant pests and diseases.

Four examples of developed qPCR methods will be discussed.

9.4.1 qPCR to Detect Phytophthora Ramorum

Phytophthora ramorum is considered an important pathogen of nursery plants in Europe and United States, where it is reported to cause twig blight of *Rhododendron*, *Viburnum*, *Camellia*, *Kalmia*, *Penis*, *Vaccinium* and other important nursery plant species. The recovery and identification of *P. ramorum* in US nurseries is increasing, where the disease often goes undetected due to the difficulty in proper diagnosis. The pathogen also causes a deadly canker disease of tanoak (*Lithocarpus densiflora*), coast live oak (*Quercus agrifolia*), California black oak (*Q. kelloggii*) and Shreve's oak (*Q. parvula* var. *shrevei*). This disease is often referred in the popular press as 'sudden oak death'. Furthermore, *P. ramorum* can cause non-lethal foliar and twig infections in numerous native hardwood and coniferous forest trees, understory shrubs and herbaceous plants in California and Oregon.

Parts of Rhododendron leaves (Fig. 9.4 left) were used for DNA extraction according Kox et al. (2007). These extracts were used as input for the assays (Fig. 9.4 right).

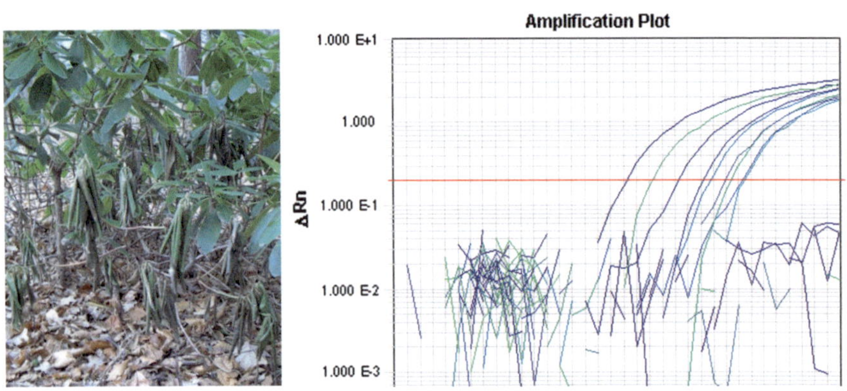

Fig. 9.4 Picture of Rhododendron leaves infected with *P. ramorum* (*left*). qPCR amplification curves for *P. ramorum* on infected plants (*right*). Samples: 19 negative (incl. 5 blanco's), 9 positive (incl. positive control)

9.4.2 qPCR to Detect Synchytrium Endobioticum

The chytridiomycete *Synchytrium endobioticum* (Schilbersky) Percival, is a plant-pathogenic fungus that causes serious infection in potato in cool and moist climates worldwide. The fungus exhibits an obligate bio-interaction with plant tissue via zoospores for infection and winter spores for survival. Upon infection, the fungus induces its host to produce tumor-like outgrowth (warts) on tubers, stolons and sprouts. Warts consist of hypertrophic host cells, each containing a single sporangium, which hosts the infectious zoospores. In senescing wart tissue, infectious zoospores are no longer produced and thick-walled resting sporangia or winter spores are formed. These spores are able to survive inter-host periods for up to 20 years.

Winter spores of *S. endobioticum* were extracted from 100 g soil samples using:

- The PPS method, which is a sieving method
- Zonal (or Hendrickx) centrifugation method.

DNA was extracted from these sieve fractions (Fig. 9.5) or zonal centrifuge fractions using the MoBio kit (Carlsbad, USA). The extraction method was optimised using the developed qPCR for *S. endobioticum* (Fig. 9.6).

9.4.3 qPCr to Detect Meloïdogyne Chitwoodi/M. Fallax

Root-knot nematodes are plant-parasitic nematodes from the genus *Meloidogyne*. They occur in soil in areas with hot climates or short winters. About 2000 plants are susceptible to infection by root-knot nematodes and they cause approximately 5% of global crop loss. Root-knot nematode larvae infect plant roots, causing the

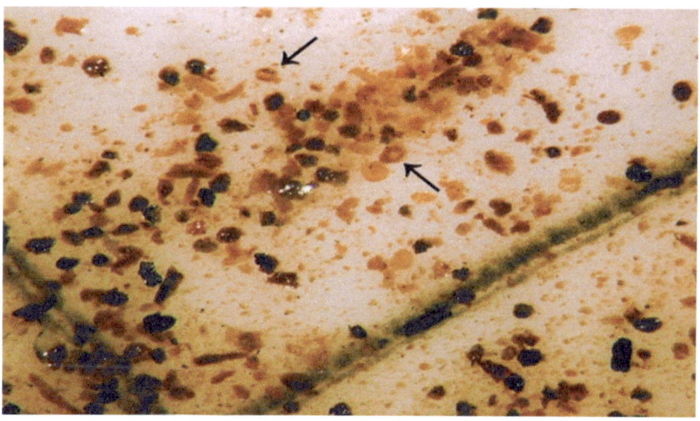

Fig. 9.5 Sieve fractions from soil containing winter spores of Synchytrium endobioticum (*arrows*)

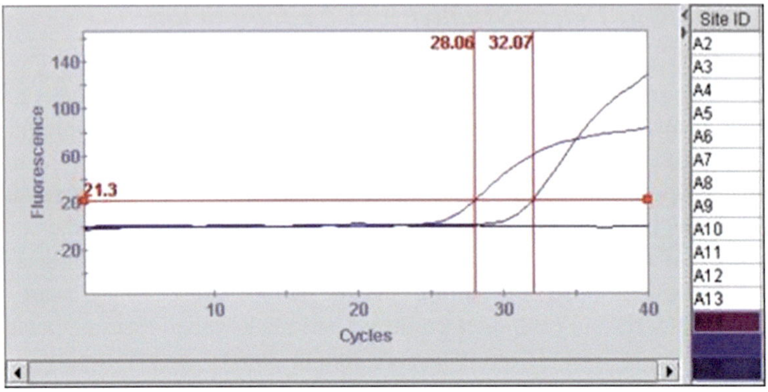

Fig. 9.6 qPCR amplification curves of two samples for the detection of Synchytrium endobioticum in soil

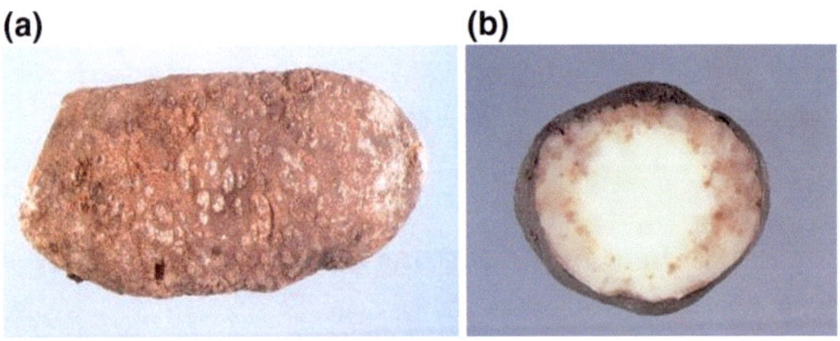

Fig. 9.7 Picture of symptoms (formation of galls and internal spots) of *M. chitwoodi/M. fallax* on potato tubers

development of root-knot galls that drain the plant's photosynthate and nutrients. Infection of young plants may be lethal, while infection of mature plants causes decreased yield. Two of the *Meloidogyne* species, *M. chitwoodi* and *M. fallax,* are on the EU quarantine list.

Potato tubers (obtained from PPS (Plant Protection Service, Wageningen, NL) (Fig. 9.7) were peeled and peelings were freeze dried, grinded and DNA was extracted using the Promega Food kit. Efficiency of DNA extraction was optimized using the qPCR method for *M. chitwoodi/M. fallax.*

9.4.4 qPCR to Detect Phyllosticta Citricarpa

Citrus black spot, caused by the fungus *Phyllosticta citricarpa* Kiely, is a foliage and fruit disease of citrus, orange, mandarin, lemon, and grapefruit (Fig. 9.8). Affected fruits become unsightly and unsuitable for the fresh fruit market.

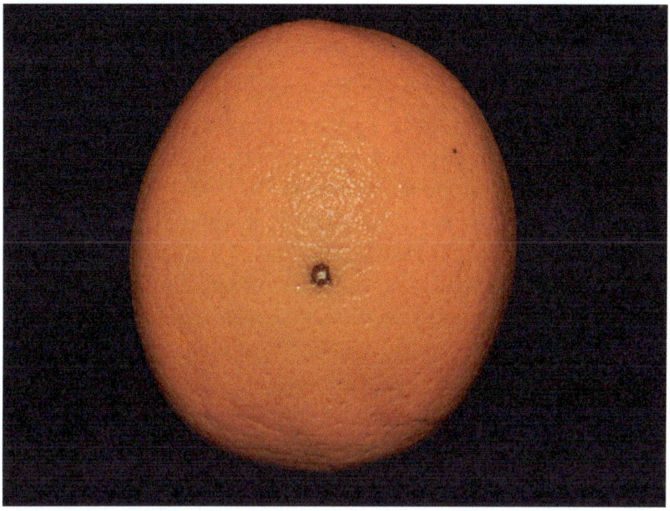

Fig. 9.8 Picture of symptoms (lesion) of *P. citricarpa* on citrus fruit

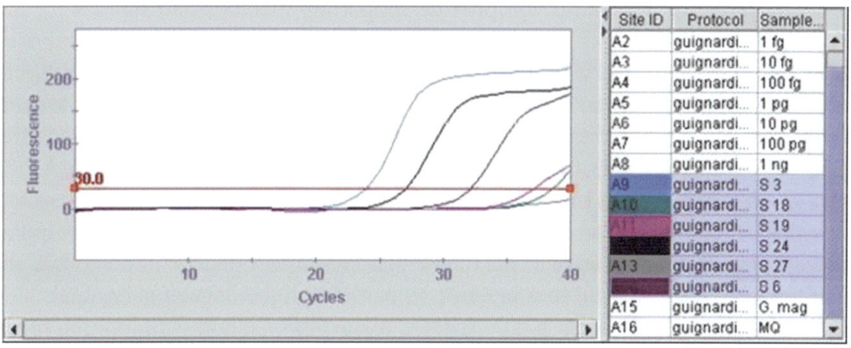

Fig. 9.9 qPCR amplification curves of practical samples (lesions) for the detection of *P. citricarpa* (formerly known as *G. citricarpa*) on citrus fruit

Premature fruit drop may occur. In areas with a warm and humid climate, losses may be substantial and intensive chemical antifungal control is required. The fungus occurs in many areas including Asia, Australia, Southern America and Southern Africa. It does not occur in the European Union (EU), where it is considered a quarantine organism.

DNA extracts, obtained according Bonants et al. (2003), were amplified with qPCR (Fig. 9.9).

9.5 EU FP7 Project TESTA

Healthy seeds are essential for high crop yields and form the basis of European and global food security. Seeds can carry a wide range of diseases and pests and, as well as spreading old problems, new problems may be introduced into the European Community countries via this route. Currently, there are opportunities to improve seed quality control by implementing novel methodologies. The TESTA project (https://secure.fera.defra.gov.uk/testa/), develops a range of novel methods to support the control of these diseases and pests, including faster and more accurate methods to assess the potency of seed transmission, economic and practical sampling approaches for the detection of low levels of those pest and diseases in large seed lots, novel and efficient generic detection methodologies, non-destructive testing methods and improved, effective and sustainable disinfection methods. Target crop and disease/pest combinations have been identified in consultation with EPPO (European Plant Protection Organisation), ISHI-ISF (International Seed Health Initiative-International Seed Federation) and ISTA (International Seed Testing Association). Outcomes from the project include a comprehensive electronic database of seed transmitted diseases and pests, validated detection methods, a validation protocol for assessing the efficacy of disinfection, as well as many key scientific publications. These will provide supporting methods and sources for the EU seed testing laboratories and plant health services. The consortium comprises experienced researchers who have been involved in previous research projects funded by the EU and national authorities, representatives of EPPO, ISTA plant health panel and ISHI working groups as well as seed testing laboratories and SMEs involved in seed production. The consortium includes a member from South Africa who is an international expert on seed production in non-EU countries, which provides insight into emerging risks. Involvement of these important players in the management of the project will guarantee that the project plans and outcomes are well-targeted and taken up in a practice so that the project legacy is ensured.

Within this EU project, a DNA/RNA extraction method was optimised in combination with qPCR methods for the detection of several tomato seed transmitted pathogens. The whole procedure for DNA extraction is described in Fig. 9.10.

The pathogens included in the study were:

- Pospiviroids: PSTVd
- Viruses: PepMV
- Bacteria: *Clavibacter michiganensis michiganensis, Xanthomonas* spp. (*X. euvesicatoria, X. vesicatoria, X. gardneri and X. perforans*) and *Pseudomonas syringae pv tomato.*

qPCRs described in the literature or available from partners were used to evaluate the whole procedure. Validation using both infected and spiked material showed that the procedure was reproducible, repeatable, and robust.

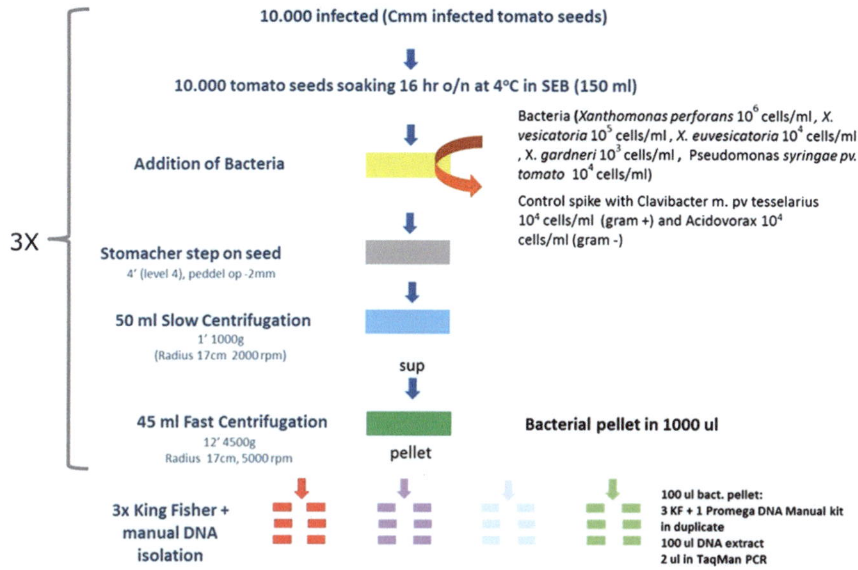

Fig. 9.10 Procedure DNA extraction experiment performed on 10,000 tomato seeds and qPCR for several tomato pathogens

9.6 Methods for Identification

Several PCR-based methods have been used for species identification of many plant pathogens. Three widely used techniques are RAPD, SSR or microsatellites and AFLPs (see Vol. 1, Sect. 2.5). Each technique has its own advantages and disadvantages. RAPD is very fast and easy to develop (because of the arbitrary sequence of the primers), but lack reproducibility. Microsatellites are specific and highly polymorphous, but they require knowledge of the genomic sequence to design specific primers and, thus, are limited primarily to economically important species. AFLP has medium reproducibility, but is labour intensive and has high operational and development costs.

9.6.1 RAPD

RAPD was used to investigate possibilities to discriminate different *Phytophthora* species (Fig. 9.11).

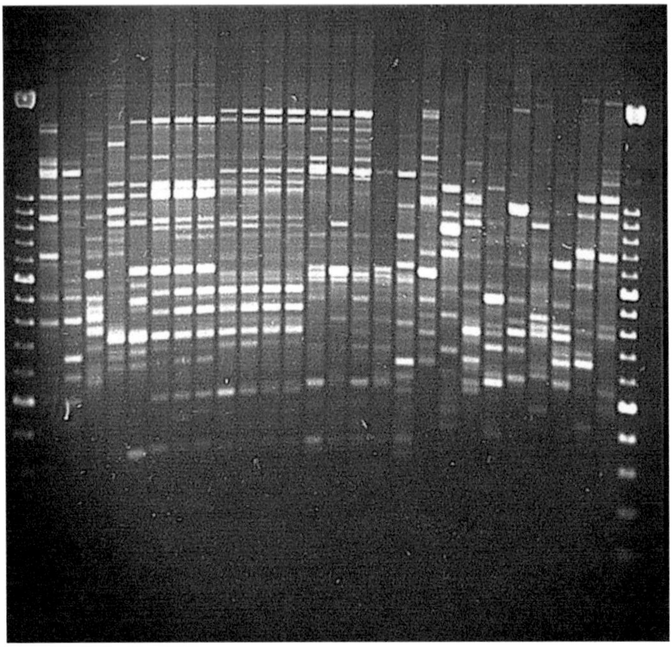

Fig. 9.11 RAPD patterns of different Phytophthora species

9.6.2 *Microsatellite Markers (SSR)*

Population genetic analysis of *Phytophthora ramorum* is critical for understanding the biology and epidemiology of this important phytopathogen. Several isolates of *P. ramorum* have been collected from several crops in Europe and the US. In the Netherlands and several European countries, *P. ramorum* is mainly found in nurseries on Rhododendron and Viburnum. Recently, for the first time, the pathogen was detected on Taxus in the Netherlands. More importantly, this pathogen was recently identified in American oak (*Quercus rubra*) in public parks in the Nijmegen area. This is the first discovery of *P. ramorum* infecting European trees. Around the same time in England also *P. ramorum* has been found in another exotic oak (*Quercus falcata*). In such cases, it is important to determine whether it is a new or old genotype, and what is the source of this infection. For tracking of the isolate in question, it is necessary to have an appropriate genotyping method. An additional advantage of this method is its applicability to study sexual crosses. Until recently, only the A1 mating type has been found in the EU and the A2 mating type in the US. However, a single isolate of the A2 mating type has now been found in Belgium and the A1 mating type was found in nurseries in the US. The occurrence of the two mating types in the same region can potentially lead to a sexually reproducing population and may generate a new source of inoculum, which might be more virulent.

Currently available methods start with pure cultures, followed by propagation and DNA isolation. However, in many cases, no pure culture can be obtained. Microsatellites are short repetitive sequences of 2 or 3 bases, so-called simple sequence repeats (Vol. 1, Chap. 2). Variation in the number of repeats can be used to study the genetic variation between isolates. Microsatellite variation can be used to give isolates a unique barcode by means of repeat PCR (Vol. 1, Sect. 5.12.6). SSR is extremely suitable to trace isolates, to study hybridization, to show differences between populations and to determine the degree of gene flow within and between populations. Microsatellite markers are potentially highly variable and also occur in humans. For this reason, SSR can also be used for genotyping humans in forensic studies.

To obtain the microsatellites, raw sequence data (445,000 reads) of *P. ramorum* were provided by the Joint Genome Institute. Our objective was to develop and utilize SSR for fingerprinting large numbers of *P. ramorum* isolates originating from different host species within Europe and the United States. Using a computer program developed within Plant Research International, 1334 potential microsatellite loci were identified. Primers were selected from over 110 flanking regions of SSRs and tested in PCR reactions to amplify repeats. Thirty-one polymorphic loci were identified and 14 primer sets were optimized for isolate genotyping. Three loci showed variation between European and US nursery isolates, although no variation was identified among isolates from the wild in the US. This information provided insight regarding the amounts of genetic variation within populations, identified new genotypes and separated isolates into two distinct lineages correlated with continental origin. All obtained 1334 useful primer combinations were included in a database. In Fig. 9.12, a part of the database is shown.

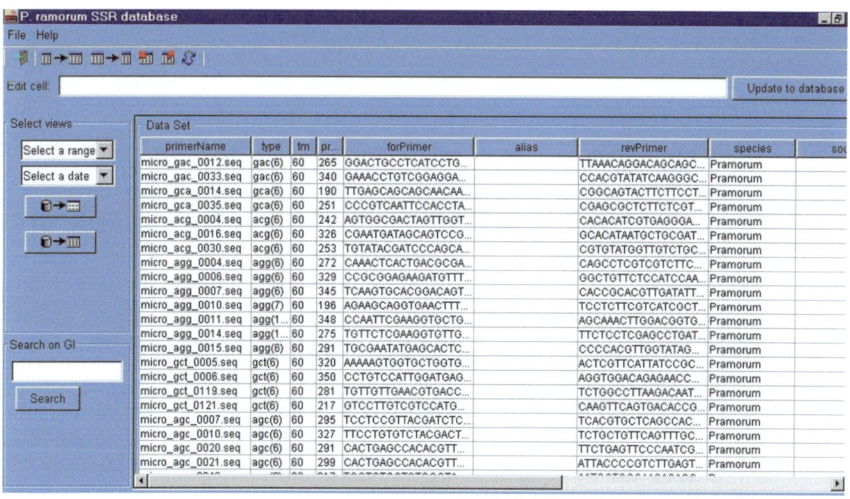

Fig. 9.12 Overview of detected primer combinations suitable for the amplification of microsatellites repeats (SSRs)

Table 9.3 Number of genome obtained and selected SSR repeats of Phytophthora ramorum for further analysis

Repeat	#	# Select	Repeat	#	# Select
AC	259	**12**	gAA	18	**3**
ACg	3	**1**	gAC	2	**1**
Ag	147	**8**	gCA	2	**1**
AgC	10	**2**	gCT	4	**1**
Agg	7	**1**	gT	226	**12**
AT	64	**3**	GTA	6	**1**
CAA	19	**1**	TAA	2	**1**
CAg	40	**3**	TAC	4	**1**
CCA	8	**2**	TAg	3	**1**
CCg	2	**1**	TCA	1	**1**
CCT	15	**2**	TCg	9	**2**
Cg	59	**6**	TgA	2	**1**
CgA	22	**3**	TgC	25	**4**
Cgg	1	**1**	Tgg	3	**1**
CgT	1	**1**	TTC	11	**2**
CT	342	**12**	TTg	7	**2**
CTg	10	**2**			
Total				1314	96

The distribution of the different primers over the different microsatellite repeats is presented in Table 9.3. Many primers to amplify the SSR were tested for polymorphisms using a standard PCR as described by Ivors et al. (2006). An example of polymorphism for several isolates of *P. ramorum* is shown in Fig. 9.13 using primer combination 18.

9.6.3 *ISSR*

Since 1994, a molecular marker technique called inter simple sequence repeat (ISSR) has been available. ISSRs are semi-arbitrary markers amplified by PCR in the presence of one primer complementary to a target microsatellite. Each band corresponds to a DNA sequence delimited by two inverted microsatellites. Like RAPD, ISSR markers are quick and easy to determine. In contrast to RAPD, reproducibility of ISSR is higher.

We tested several ISSR primers (Table 9.4) to study genetic variation between US and EU isolates of *P. ramorum*. Results of PCR amplification with primer P are shown in Fig. 9.14. Analysis of multiple polymorphic PCR primers resulted in a dendrogram as shown in Fig. 9.15, which clearly indicate that the US population is distinct from the EU population of *P. ramorum*.

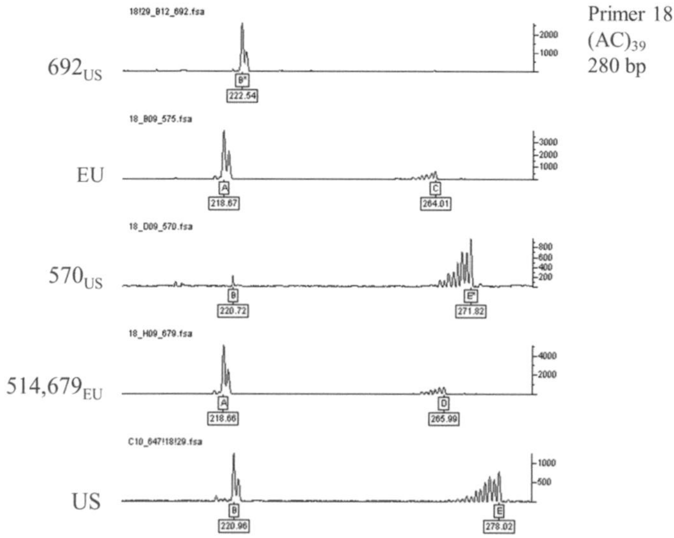

Fig. 9.13 Different SSR profiles of several isolates of *P. ramorum* from EU and US obtained for primer combination 18

Table 9.4 ISSR primers used in *P. ramorum* research. Ambuigity bases are described in Chap. 5

A: NDB(CA)7-C	B: NDV(CT)8	C: (GT)9-C	D: (GT)9-A
E: DHB(CGA)5	F: (GA)9-C	G: DDB(CCA)5	H: (GA)9-A
I: BDB(ACA)5	J: (CT)9DV	K: (CA)9-DB	L: (CT)9HV
M: (GA)9HB	N: (GAG)6HB	O: (GTG)6HV	P: (CTC)6DV
Q: (CAC)6DB	R: NDV(CT)8	S: VBVH(TATG)4	T: (TATG)4SS

9.6.4 AFLP

AFLP (Vol. 1, Sect. 5.12.6) is a powerful technique to identify and to study genetic variation in fungi. *Rhizoctonia solani* is a fungal pathogen of sugar beet. Isolates from different sugar beet fields were collected and analysed by AFLP DNA fingerprinting to study genetic variation. Figure 9.16 shows that the genetic variation of isolates within one field is zero while there is some variation between fields.

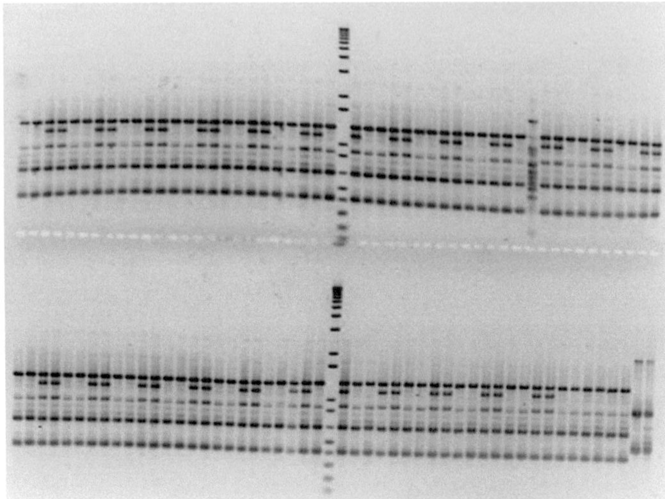

Fig. 9.14 Gelectrophoresis of ISSR PCR on 96 DNA extracts of EU and US *P. ramorum* isolates amplified with one ISSR primerset, showing distinct bands to discriminate EU from US isolates of *P. ramorum*

9.7 Detection EAB

In the USA, the Emerald Ash Borer (EAB) (*Agrilus planipennis*) is a very important insect pest organism (EU Q-organism) that causes great harm in ash trees and other host plants. Recently, this species was found in Moscow (Russia), being the first reporting on the European continent. In the juvenile stage, this beetle is barely distinguishable from native species when using conventional methods. A fast, accessible, and reliable method to recognize *A. planipennis* larvae at an early stage is highly wanted. The aim of the project was to develop a specific identification method that can be applied to (parts of) an individual insect (leg, head, larva, whole insect). For this, *A. planipennis* material from USA and Asia (China) and related species from Western Europe were collected and the COX-1 (cytochrome oxidase-1) gene was sequenced. Based upon these sequence data and on sequence data available from the Genbank database, all sequences were aligned and a dendrogram was made to illustrate genetic relatedness. This dendrogram of COX-1-sequences of many *Agrilus* species (Fig. 9.17) shows a clear distinction between EAB and other *Agrilus* species. A specific qPCR, based on these sequence differences, was tested on DNA isolated from a leg, head, pieces of larvae and whole insects (Fig. 9.18).

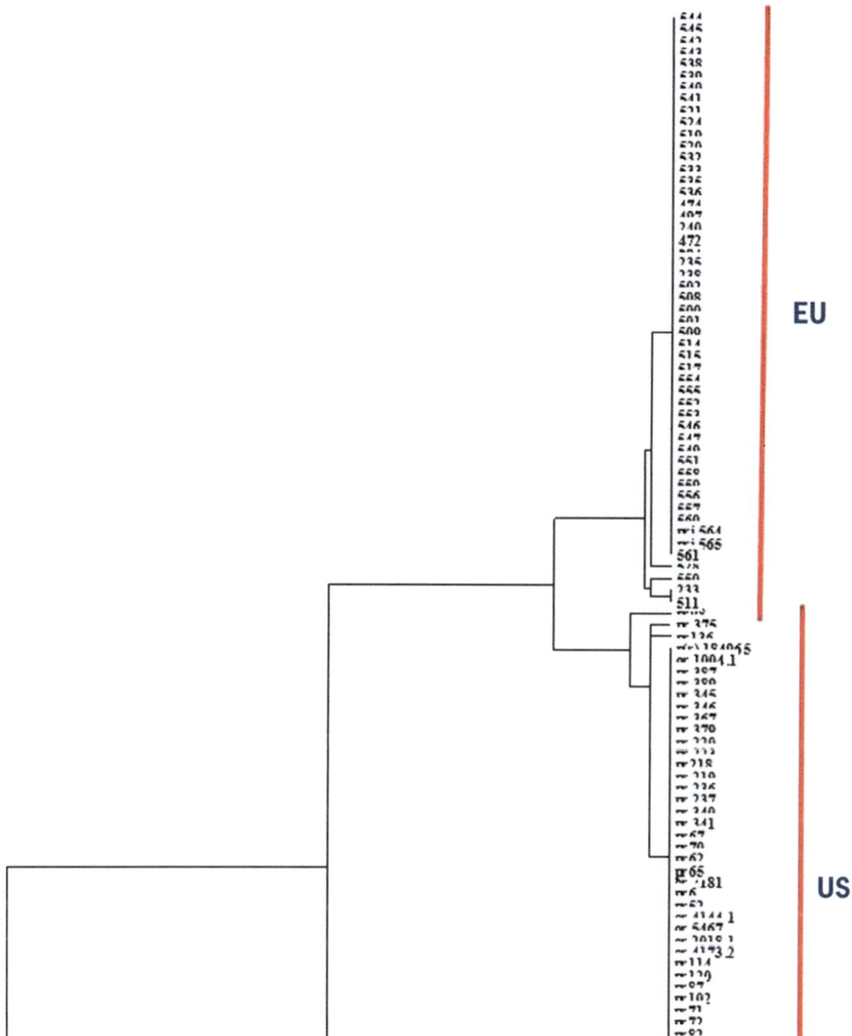

Fig. 9.15 Dendrogram of ISSR PCR electrophoresis patterns using multiple PCR primer sets on DNA extracts of EU and US *P. ramorum* isolates

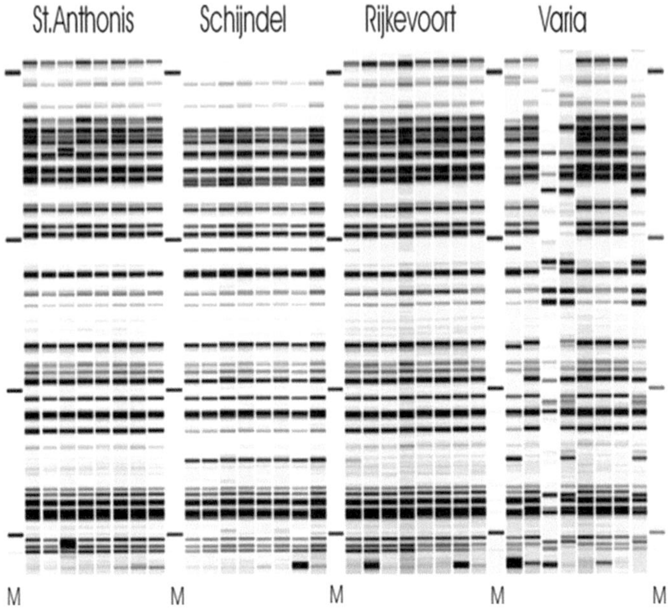

Fig. 9.16 AFLP patterns of Rhizoctoni solani isolates obtained from 4 different sugar beets fields in the Netherlands

9.8 DNA Barcoding in FP7 EU Project QBOL and the Database Q-Bank

Development of accurate identification tools for plant pathogens and pests is vital to support European Plant Health Policies. For the EU project QBOL (Bonants et al. 2010) Council Directive 2000/29/EC is important, listing around 300 organisms for which protective measures against introduction into and their spread within the Community needs to be taken. Those threats are now greater than ever because of the increase in volumes, commodity types and origins of trade, the introduction of new crops, the continued expansion of the EU and the impact of climate change. Currently, identifying pathogens (in particular new emerging agents) requires staff with specialized skills in all disciplines (mycology, virology, bacteriology, etc.). This is only possible in large centralized laboratory facilities. Especially now that taxonomy, phytopathology and other fields that are vital for sustaining sound public policy on phytosanitary issues are "threatened with extinction". In the past, identification was performed with classical tools. Recently, this expertise is diminishing and replaced by molecular identification tools. These molecular techniques often require other specialist skills to perform, are more amenable for routine purposes and can be used for a whole range of different target organisms. Recently, DNA barcoding has arisen as a robust and standardized approach for species identification. A DNA barcode is a short region of the DNA, which can be used to identify a

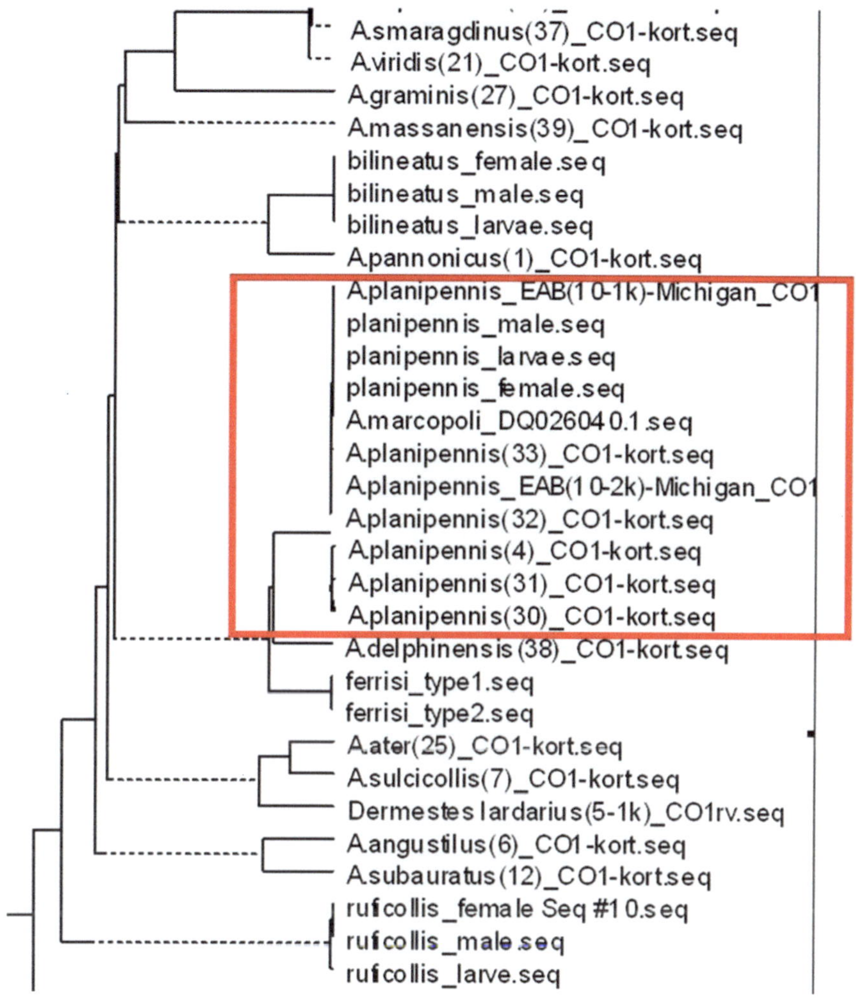

Fig. 9.17 Dendrogram of COX-1 (cytochrome oxidase-1) sequences of many Agrilus insects. Asterix indicate sequences obtained from insects from the USA. Circle indicates *A. planipennis* sequences

species. QBOL (www.qbol.org) made DNA barcoding available for plant health diagnostics and focused on strengthening the link between traditional and molecular taxonomy as a sustainable diagnostic resource. Within QBOL, collections harboring plant pathogenic Q-organisms were made available. Informative genes from selected species on the EU Directive and EPPO lists have been DNA barcoded from vouchered specimens and the sequences, together with taxonomic features, have been included in a new internet-based database system: Q-bank: www.q-bank.eu (Bonants et al. 2013). This database is freely accessible via Internet. Tools have

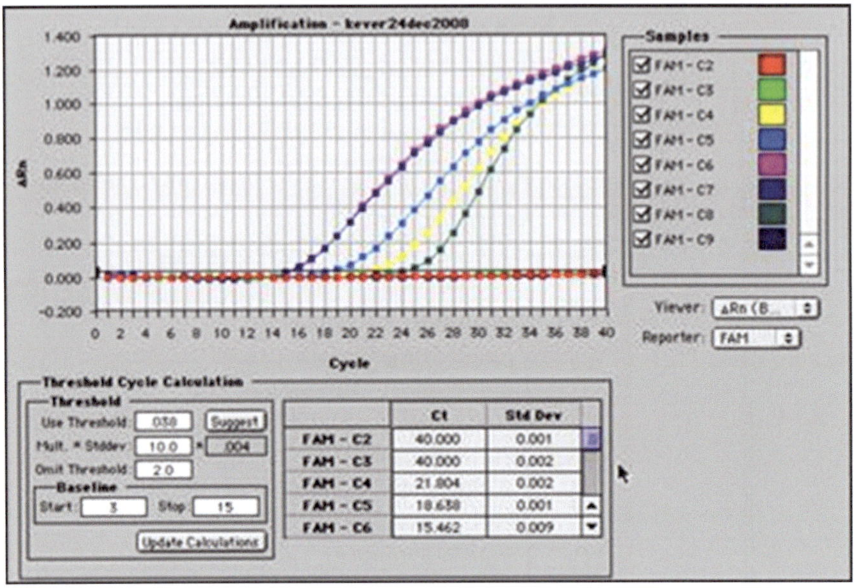

Fig. 9.18 qPCR amplification curves of DNA extracts isolated from a leg, head, piece of a larvae and from a whole insect of *A. planipennis*

been provided on how to search the database and to perform BLAST analysis or even multilocus and/or polyphasic identification.

To ensure robustness of the developed procedures for the use in a distributed network of laboratories across Europe, a validation procedure on both the protocols and the database has been performed across worldwide partners. Q-bank was first developed in a Dutch FES project financed by the ministry of Economic Affairs. This project (2006–2010) aimed to strengthen the infrastructure on plant health. During this project and in some other small projects from the NPPOs, universities, and institutes, more data were incorporated in the seven databases of Q-bank. Curators from different plant protection organizations, universities and institutes took care that included data met certain standards. Important to mention is that all data were retrieved from specimens, which were available in collections.

Within the QBOL project we were able to develop DNA barcodes for many plant pathogenic Q-organisms that are present on the EU Directive and EPPO list and closely related species. Protocols for DNA/RNA extraction, generic amplification of the barcoding region and sequence analysis were written and included in the molecular decision schemes. These schemes were produced by work packages 2–7 and allow end-users to use the protocols for correct identification of the q—organisms. In Table 9.5 the total number of sequences obtained within the QBOL project is depicted.

For each of the different groups of organisms a “molecular decision scheme” was designed. An example for bacteria is shown in Fig. 9.19. The scheme clearly shows

Table 9.5 Number of obtained barcoding sequences for the identification of different groups of organisms within the different work packages of the QBOL project

QBOL	Sequences obtained		3-10-2012
		# Sequences	Remark
WP2	FUNGI	6898	8 loci
WP3	ARTHROPODS	5300	2 loci
WP4	BACTERIA	3667	20 loci
WP5	NEMATODES	1683	6 loci
WP6	VIRUSES	46	Whole genome seq
WP7	PHYTOPLASMS	472	3 loci

which protocols have to be followed and which genes have to be sequenced. When participants follow this scheme, it is ensured that inspection for the presence of q-organisms is performed in the correct way.

9.9 Future Developments

9.9.1 Multiplex Approach for Multiple Pathogen Detection

An interesting platform for multiplex detection is the Luminex system (Fig. 9.20). The Luminex system uses 100 different coloured beads (about six microns in size), all of which can detect their own pathogen-specific target(s). These beads can, for instance, be loaded with specific antibodies that react to a protein of a particular pathogen using a technology, which is called xMAP. Nucleic acids can be coupled to the beads as well, enabling multiplex nucleic acid detection. This is called xTAG.

The Luminex device uses two laser beams. The first laser detects and identifies the bead which passes through the laser beam. The second laser detects the presence or absence of coupled antibody or DNA fragment on that bead. The system can be expanded easily to detect more pathogens since 100 different beads are available.

9.9.2 On-Site

Another development is the On-site or Point-of-Care detection method, which can be performed in the field. Wageningen University and Research Centre (WUR), together with international partners, has been working on the development of a new and rapid method for on-site detection of different plant pathogens. This detection method is based on reproduction of specific DNA or RNA fragments at a constant temperature of 65 °C while using the so-called 'Loop mediated AMPlification' (LAMP) technology (see Vol. 1, Chap. 7). The big advantage of this approach is that the DNA/RNA in the (plant) sample does not need to be purified, as is the case

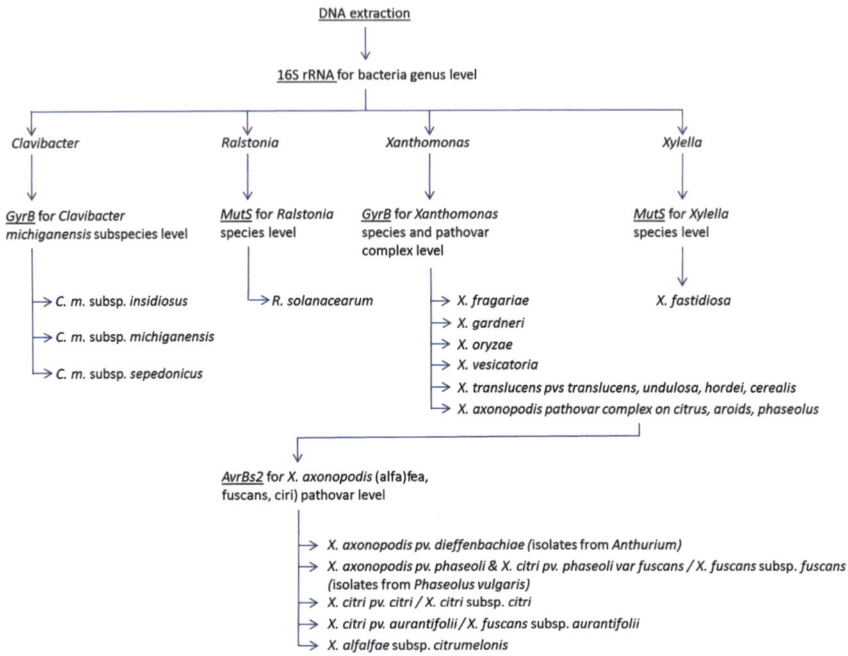

Fig. 9.19 Molecular Decision scheme for DNA barcoding of quarantine bacteria, as obtained within work package 4 of the QBOL project

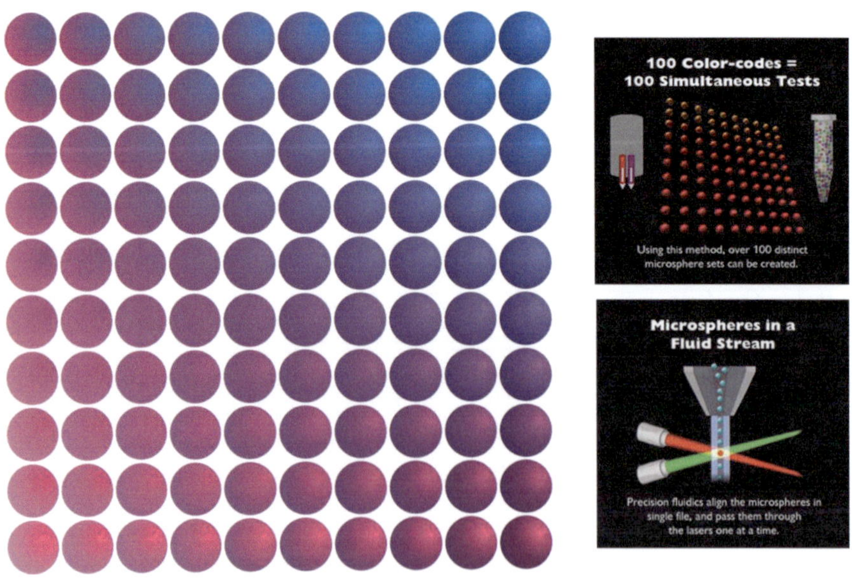

Fig. 9.20 Luminex system for multiplex detection

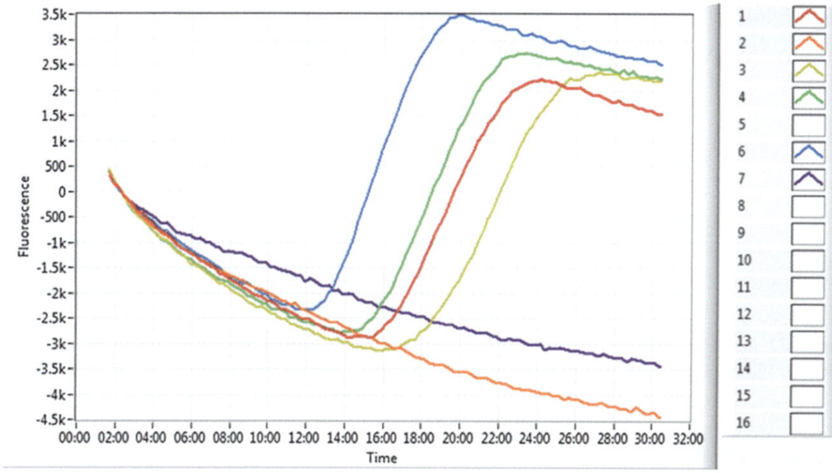

Fig. 9.21 On-site detection of Phyllosticta citricarpa using LAMP and the Genie III on 4 different lesions on citrus fruit

for most molecular assays. After a short and simple preparation this method enables on-site determination of the presence and/or identity of a specific pathogen or insect, within 10 min. This is done with a Genie II or III instrument (Optigene, Horsham UK) and without special training in molecular biology. LAMP applications for many plant pathogens have been developed and published. One example is the use of LAMP for detection of e.g. *Phyllosticta citricarpa*, a fungal pathogen of citrus (Fig. 9.21).

9.9.3 Vital Versus Dead

All amplification methods use DNA and/or RNA, but very often the question remains: is the detected organism vital or alive? Several approaches have been studied to address this question. Intact organisms such as bacteria, fungi and parasites have an intact membrane. After death, this membrane disintegrates and certain chemical compounds can enter the organism. One of these compounds is propidium monoazide (PMA). PMA is a membrane-impermeant dye that selectively penetrates cells with compromised membranes (read: dead cells). Once inside the cell, PMA intercalates into the DNA and can covalently cross-link to it, which strongly inhibits PCR amplification. In this way, living cells (no bound PMA, PCR is possible) can be discriminated from dead cells (bound PMA, no PCR is possible). An example of this approach is shown in Fig. 9.22, in which a qPCR is performed on dead and living *Xanthomonas* bacteria. A shift in Cq-value to the right is observed between dead and living bacteria after PMA treatment.

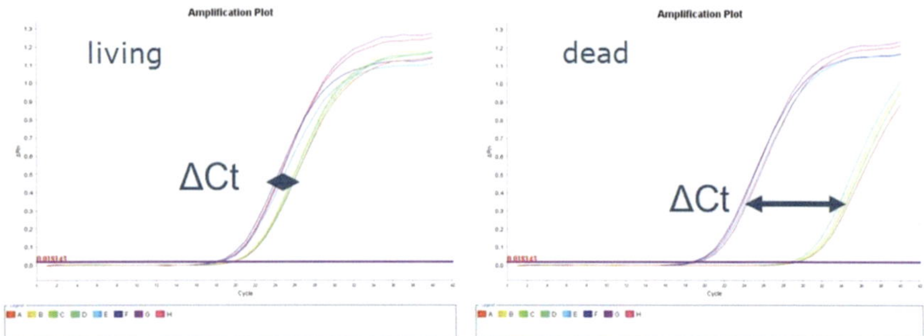

Fig. 9.22 qPCR of living (*left*) and dead (*right*) Xanthomonas bacteria after treatment with and without PMA

9.9.4 Next Generation Sequencing (NGS)

In the last decade, traditional Sanger Sequencing is being replaced by Next Generation Sequencing (NGS) platforms (Vol. 1, Chap. 8). NGS, also known as high-throughput sequencing, is the catch-all term used to describe a number of different modern sequencing technologies, each with their own characteristics.

- Illumina (Solexa) sequencing
- Roche 454 sequencing
- Ion torrent: Proton/PGM sequencing
- SOLiD sequencing
- PacBio sequencing.

These recent technologies allow high-throughput sequencing of DNA and RNA much faster and cheaper than the previously used Sanger sequencing procedure. As such, NGS has revolutionized the study of genomics and molecular biology. The four main advantages of NGS over classical Sanger sequencing are: speed, cost, sample size, and accuracy. However, analysis of the obtained sequences needs extensive bioinformatics.

Several platforms were used to evaluate the applicability of NGS for the analysis of different RNA and DNA extracts, such as extracts from lily to study viruses, extracts from potato to study bacterial populations and extracts from tomato seeds to study which pathogens are seed-transmittable. In these studies, whole genome sequencing of plant pathogenic viruses, bacteria and fungi was performed and results showed that presence of plant pathogens could be proven by NGS analysis.

Most of the gene sequences used nowadays in plant health diagnostics are based on housekeeping or ribosomal genes. Future developments in diagnostics in plant health will focus more on functional markers, which are linked to functional genes. Those genes play an important role in pathogenesis.

Microfluidics and nanosensors already play a role in human health diagnostics. Once their prices drop, they certainly will also become part of the diagnostics in plant health.

9.10 Conclusions

In the last two decades, molecular methods, based on DNA/RNA, have become an important part of diagnostics and detection methods. Furthermore, many new applications and techniques will be developed and will be put on the market during the coming years. Although molecular technologies for the detection and identification of plant pathogens are already (semi-) routinely in use in a number of places (and are even available commercially), it is not always clear whether a full validation for these tests has been performed. Since each method/pathogen combination has its own characteristics, proper validation and comparison to existing methods is mandatory. This can be done by the Plant Protection Service in cooperation with the practical laboratories. Performance criteria playing an important role are technical and diagnostic sensitivity and specificity, robustness and LOD (see Vol. 1, Chap. 6). These criteria not only depend on the extraction method used, but also highly on the type of plant material (roots, leaf, wood, flowers, etc.). Experience has also shown that not every new (molecular) analysis can be introduced as standard in every laboratory. An implementation phase as well as a much higher level of quality assurance is required.

A variety of identification and detection methods have been discussed in this chapter. Methods to test plants and plant products for the presence and absence of a highly diverse group of plant pathogens and pests have been summarized. DNA and RNA sequences of the target organisms or closely relates species need to be determined. Based on sequence differences, specific tests can be designed. More and more specific assays will become available to be used in the laboratory or in the field on-site. Test for one (monoplex) or several (multiplex) pathogens will become more sensitive and cheaper. Also quantification, sampling strategies, and linkage with decision support systems will be incorporated in future.

It is very important to have access to strain collections, which contain the necessary isolates, are pure and very well characterized. Also expertise in phytopathology and molecular biology are indispensable for the development of detection and identification methods.

Acknowledgements This chapter has been written in collaboration with many national colleagues at Wageningen UR, the Dutch Inspection Services (NAK, Naktuinbouw, BKD and KCB) and the Dutch Plant Protection Service (NVWA).

Much of the work has been part of many EU projects (Portcheck, QBOL, Q-bank, TESTA), in which many international contacts were used.

Key References

Aveskamp MM, De Gruyter J, Crous PW. Biology and recent developments in the systematics of Phoma, a complex genus of major quarantine significance. Fungal Divers. 2008;31:1–18.

Bonants PJM, Carroll GC, de Weerdt M, van Brouwershaven IR, Baayen RP. Development and validation of a fast PCR-based detection method for pathogenic isolates of the citrus black spot fungus, Guignardia citricarpa. Eur J Plant Pathol. 2003;109:503–13.

Bonants P, Edema M, Robert V. Q-bank, a database with information for identification of plant quarantine plant pest and diseases. EPPO Bulletin. 2013;43(2):211–5.

Bonants P, Groenewald E, Rasplus JY, Maes M, de Vos P, Frey J, Boonham N, Nicolaisen M, Bertacini A, Robert V, Barker I, Kox L, Ravnikar M, Tomankova K, Caffier D, Li M, Armstrong K, Freitas-Astua J, Stefani E, Cubero J, Mostert L. QBOL: a new EU project focusing on DNA barcoding of Quarantine organisms. Bulletin OEPP/EPPO Bulletin. 2010;40:1–5.

Ivors K, Garbelotto M, De Vries I, Ruyter-Spira C, te Lintel B, Hekkert NR, Bonants P. Microsatellite markers identify three lineages of *Phytophthora ramorum* in US nurseries, yet single lineages in US forest and European nursery populations. Mol Ecol. 2006;15:1493–505.

Kox L, Brouwershaven IV, van den Vossenberg B, van den Beld H, Bonants P, de Gruyter J. Diagnostic values and utility of immunological, morphological, and molecular methods for in planta detection of Phytophthora ramorum. Phytopathol. 2007;97:1119–29.

van Gent-Pelzer MPE, van Brouwershaven IR, Kox LFF, Bonants PJM. A Taqman PCR method for routine diagnosis of the quarantine fungus *Guignardia citricarpa*. J Phytopathol. 2007;155:357–63.

van Gent-Pelzer MPE, Krijger M, Bonants PJM. Improved real-time PCR assay for detection of the quarantine potato pathogen, *Synchytrium endobioticum*, in zonal centrifuge extracts from soil and in plants. Eur J Plant Pathol. 2009;126(1):129–33.

Chapter 10
Whole Genome Sequencing in Food Outbreak Investigation and Microbial Risk Analysis

Henk Aarts and Eelco Franz

Abstract Next Generation Sequencing (NGS) is a huge technological advance in the molecular typing of micro-organisms. This chapter describes the use of NGS in food-borne outbreak investigation, source attribution and also describes the first steps in using Whole Genome Sequencing (WGS) data in molecular risk assessment. The rapidly decreasing costs and operational time make that WGS will likely replace currently used molecular typing methods, such as MLST, MLVA and PFGE, in outbreak investigation and source attribution in the near future. Because of the superior level of resolution and because all genetic information (including virulence and antimicrobial resistance genes) of the organisms can be revealed, WGS can be considered as the ultimate "all-in-one" typing method.

Keywords Food-borne pathogens · Whole genome sequencing · Outbreaks · Typing · Microbial risk analysis

10.1 Introduction

Within the field of genomics, Whole Genome Sequencing (WGS) is becoming "the" method of choice for addressing foodborne outbreaks and for assessing the potential risks of pathogens. Next Generation Sequencing (NGS) platforms produce a large volume of data in a relatively short time and have turned out to be a powerful tool for WGS of viral and bacterial genomes. This was demonstrated

H. Aarts (✉)
Centre for Zoonoses and Environmental Microbiology (Z&O),
National Institute for Public Health and the Environment (RIVM),
Bilthoven, The Netherlands
e-mail: henk.aarts@rivm.nl

E. Franz
Centre for Infectious Diseases, Epidemiology and Surveillance,
National Institute for Public Health en the Environment (RIVM),
Bilthoven, The Netherlands

E. van Pelt-Verkuil et al. (eds.), *Molecular Diagnostics*,
DOI 10.1007/978-981-10-4511-0_10

during the recent entero-haemorrhagic *Escherichia coli* (EHEC) outbreak in Europe. Sequencing of the outbreak strain of *E. coli* O104 during the outbreak in Germany in 2011 took only 3 days, with the first assembly being released two days after completion of sequencing. Within one week, the data became available, showing a novel entero-aggregative *E. coli* O104 variant, which had acquired a prophage encoding Shiga toxin 2. The concept behind NGS technology is not different from Sanger sequencing performed in combination with capillary electrophoresis. However, the huge difference is that the number of reactions, which are run in parallel (1×10^6 times), is hugely increased in NGS, enabling the rapid sequencing of an entire genome of a microorganism in a single run at (relatively) low cost.

10.2 WGS in Relation to Food Safety

Because of its superior discriminatory power, WGS has huge potential for the detection and investigation of outbreaks, source attribution (i.e. the relative contribution of different reservoirs to the disease burden) and risk assessment. WGS will likely replace serological and molecular typing methods, such as serotyping, multilocus sequence typing (MLST) and probably also MLVA (multilocus variable number of tandem repeats analysis) and Pulse Field Gel Electrophoresis (PFGE). Furthermore, the characterization of the genetic content of pathogens will enable the identification of health related properties like virulence and antibiotic resistance. WGS will also provide the necessary genomic information for the development of diagnostic tests based on significant markers and for the detection of new variants. However, several challenges have still to be solved in order to make full use of the potential of WGS for food safety (EFSA 2013). These include storage and sharing of the data, harmonization of bioinformatics pipelines and standardized nomenclature and interpretation.

10.2.1 Molecular Typing Methods

Methods used for the characterization of bacterial or viral strains below the (sub-) species level, are by definition named “typing methods”. Important performance criteria for these methods are:

(i) Typeability: the proportion of strains that are assigned to a type by the typing method
(ii) Discriminatory power: the likelihood that strains can be distinguished by the typing method
(iii) Repeatability: the variation in the correct assignment performed under the same conditions (person, system, time etc.)

(iv) Reproducibility: the variation in the correct assignment performed under different conditions (person, system, time etc.)
(v) Stability: the ability of a typing system to recognize the clonal relatedness of strains derived from a common ancestor strain.

Typing methods can be divided in methods based on phenotypic properties (for example serotyping, phage typing, biotyping, antimicrobial resistance and protein profiling) and in methods based on the genetic content, such as DNA fingerprinting (PFGE, RAPD, AFLP, ERIC, MLVA) and DNA sequencing (MLST, full genome) techniques. All these methods have their own advantages and disadvantages. PFGE, for instance, is expensive and time consuming (4–5 days). On the other hand, PFGE is a method that can be used for all bacteria, that has a high discriminative power and that shows a high level of typeability. Table 10.1 shows a comparison (discriminative power, number of markers, reproducibility etc.) of commonly used bacterial typing techniques, which also includes the cost per isolate analysis. Sabat et al. (2013) have created a comprehensive overview, including the pros and cons, of the existing molecular typing methods that are used for outbreak investigations and epidemiological surveillance. The most advanced molecular typing method is WGS. For this method, the genome has to be fragmented and subsequently sequenced. The latter is done in specialized sequencers like the 454 Life Sciences (GS FLX Titanium) from Roche, the Ion Torrent & ABI Solid from Life Technologies, the HiSeq, MiSeq and GenomeAnalyzer from Illumina or the PacBio RS from Pacific Biosciences. In using WGS/NGS, huge amounts of data are generated and analysis of these data requires massive bioinformatics support and storage capacity (see also Volume 1, Chaps. 8 and 9).

An alternative upcoming method for microbial typing is Matrix-Assisted Laser Desorption/Ionization-Time Of Flight (MALDI-TOF). This mass spectrometry method has great advantages in that it is fast, easy to use, high through-put and low cost. However, a limitation of MALDI-TOF is that it does not reliably identify isolates at the subspecies and clonal level, it does not give sufficient information on different virulence and resistance genes and it is difficult to standardize in a way that enables the global exchange of data.

10.2.2 Genomic Subtyping of STEC O157

Shiga-toxin producing *E. coli* (STEC) is a zoonotic pathogen that causes diarrheal disease in humans and is of public health concern because of its ability to cause outbreaks and severe disease, such as haemorrhagic colitis (HC) or haemolytic-uremic syndrome (HUS). Numerous cases of HC and HUS have been attributed to EHEC serotype O157:H7 strains. It can be considered paradoxical that the number of human STEC O157 infections is rather low for a pathogen with a low infectious dose and a relatively high prevalence in the ruminant reservoir. This could, at least partly, be explained if only a subset of STEC O157 isolates present in

Table 10.1 Comparison of most common bacterial typing techniques (adapted from Foxman et al. 2005) by Pérez-Losada et al. (2013)

Typing method	Method description	No of markers	Temporal scale	Variation source	Discriminatory power	Reproducibility	Equipment/Time	Equipment/consumables-Reaction costs (per isolate)	Available databases
MLST	PCR amplification of housekeeping genes to create an allelic profile	7	Macroepidemiological Microepidemiological	DNA sequence	Moderate to high	High	Thermal cycler/Moderate	\$30K–45K High–\$80	Pubmist.org www.mist.net mlst.ucc.ie www.pasteur.fr/mlst
MLEE	Phenotypic characterization of the elcotrophoretic mobility of housekeeping enzymes	10–20	Macroepidemiological Microepidemiological	Electrophoretic mobility	Moderate	Moderate	Gel box, switching unit cooler, power supply/Moderate	\$10K–20K Moderate–\$20	NA
PFGE	Comparison of large genomic DNA fragments after digestion with rare restriction enzyme	NA	Microepidemiological	Banding pattern	Moderate to high	High	Gel box, switching unit cooler, power supply/High	\$10K–20K Moderate–\$22	NA
AFLP	Digestion of genomic DNA with two restriction enzymes, ligation of restriction fragments and selective amplification	NA	Microepidemiological	Banding pattern	Moderate to high	Low	Thermal cycler/Moderate	\$8K–12K Moderate–\$20	NA
MLVA	PCR amplification of VNTR loci followed by sizing of the PCR products to create an allelic profile	10–80	Microepidemiological	DNA sequence	Moderate to high	High	Thermal cycler/Low	\$30–45K Moderate–\$20	Minisatellites.u-psusd.fr www.mlva.net www.pasteur.fr/mlst
HRM	PCR amplification followed by characterization of amplicon melting curves	NA	Macroepidemiological Microepidemiological	Melting temperature	High	High	Real time thermal cycler/Very low[a]	\$30K–45K Very low[a]	NA
RFLP	Digestion of genomic DNA with restriction enzymes to produce	NA	Microepidemiological	Banding pattern	Low	Low	Southern transfer/High	\$8K–12K Low–\$14	NA

(continued)

Table 10.1 (continued)

Typing method	Method description	No of markers	Temporal scale	Variation source	Discriminatory power	Reproducibility	Equipment/Time	Equipment/consumables-Reaction costs (per isolate)	Available databases
	multiple short restriction fragments								
rMLST	PCR amplification of *rps* genes to creat an allelic profile	53	Macroepidemiological Microepidemiological	DNA sequence	High	High	Thermal cycler/Moderate	\$30K–45K High–\$600 (if WGS is not needed)	http://pubmlst.org/software/database/bigsdb/
Pan-genome	Detection of similarities/differences in the pangenomic or distributed genes	>1000	Macroepidemiological Microepidemiological	Presence/absence of genes	High	High	NGS platforms or Microarrays/Moderate to high	\$80K–130K Very high–\$1K–20K per run depending on the NGS platform used	www.francisella.org

[a]If new melting profiles are not detected

the bovine reservoir are characterized by unique transmissibility and/or virulence characteristics. Distinguishing STEC O157 isolates that constitute a high risk to human health from isolates that have a lower association with clinical symptoms is an important aspect of risk-based monitoring and surveillance.

Franz et al. (2012) genetically characterized and compared bovine, food, and human clinical STEC O157 isolates from the Netherlands using various genotyping methods and multivariate statistics. The goal was to determine whether different genotypes occur at different frequencies among isolates from these different sources and to identify the most differentiating genetic features of the isolates from these sources. A set of 73 bovine, 29 food, and 85 human clinical isolates was used. The results clearly showed non-random distributions of genotypes among bovine, food, and human clinical isolates. In addition, comparison with published data revealed significant differences in the distribution of genetic lineages among bovine and human isolates between the Netherlands versus North America and Japan. As a follow-up to get more insight in the genomic and phenotypic differences between STEC O157 isolates from different geographic regions, strains from The Netherlands (n = 39) (low *E. coli* O157 disease incidence) and Scotland (n = 145) (high *E. coli* O157 disease incidence) were subjected to whole genome sequencing. Significant differences in genetic distance were observed between isolates from different countries. It was therefore concluded that after successful transmission, which occurred several times, local genetic variation occurs, resulting in a geographical patchwork of phylogeographical clades.

10.2.3 Whole Genome Sequencing of Food Borne Pathogens

Sequencing the whole genome of a pathogen results in huge amounts of data, which consists of many individual reads. In recent years, we have seen a rapid increase in the raw daily output of the different platforms and a decline in turnaround times for bacterial genomes and in cost per Mb assembled sequence.

The most common way to analyse WGS data is the reference-based single-nucleotide-polymorphism (SNP) approach. Here, the reads are mapped to a complete and reliable sequenced reference genome, which should be as closely related as possible to the strains under investigation. Subsequently, SNPs are called in the reads that map to the reference genome. Drawbacks of this reference method are that a closely related reference genome is necessary to map as much reads as possible and that a sequences absent in the reference genome will not be mapped. The latter may introduce a bias when the strains under investigation are diverse. For these reasons, researchers often choose the approach of reference-free '*de novo*' assembly. Hereby, reads are concatenated into longer stretches of reads, so-called contigs. This is done by specific algorithms and based on overlaps between reads. In turn, overlapping contigs can be assembled into scaffolds. A final step may be the closure of the entire genome, but this is often difficult due to gaps in the genome and the presence of repetitive elements. The quality of this de novo assembly will increase with the use of sequence platforms that produce longer reads, since these leave less uncertainty in the assembly procedure. The assembled genomes can subsequently be used for SNP detection, phylogeny (based on all genetic information, rather than the SNPs only, as with the mapping approach) or presence/absence detection of specific genes.

An alternative to the reference-based or reference-free SNP approach is the gene-by-gene approach. In this case, allelic variation (which can be caused by SNPs, recombination, and/or insertion/deletion events) in pre-defined loci are identified. This approach is already applied in the classical MLST, where Sanger sequencing determines the sequence variation of a predefined set of household genes. The resulting allelic variations are translated into a numeric nomenclature, which is curated. With WGS, this classical MLST can be performed in silico, but it can also be extended to the all known loci in the core-genome (i.e. the part of the genome shared by all strains under investigation) or even to the pan-genome (i.e. unique genes). The development of a core-genome or genome-wide MLST scheme, however, does require a closed reference genome and preferably a set of query genomes (which can be in contig or scaffold stage) in order to define a more stable scheme.

The reads of a single bacterial genome, at a level of a 100-fold coverage will take 200 MB of disk space. Handling the information of a large number of strains requires a complex IT infrastructure, which generates computational challenges that must be addressed by specialized software. Many platforms incorporate superseded versions of algorithms. Various algorithms exist, for instance for '*de novo*' assembly and RNA-sequencing. Besides commercial packages, there are also freely available "open-resource" algorithms from academic institutions. The available algorithms make it possible to assemble genomes in many instances, meaning that the assembly can be performed using the existing computer resources in the laboratory. For instance, an *E. coli* genome can be assembled within a time frame of 15 min on a 32 bit Windows desktop computer with 32 GB of RAM.

Recently, several bioinformatics tools have been published, such as "RESfinder". This software package is developed by the Technical University of Denmark (DTU) and subtracts antibiotic resistance gene (or related sequences) information from whole genome sequences. They showed that there was a good concordance between the "in silico" data and data obtained from phenotypic analysis (personal communication). Besides RESfinder, a number of other software packages, like PathogenFinder, KmerFinder, VirulenceFinder, are available on their website (http://www.genomicepidemiology.org). Although good concordance between the detected genes and phenotypic properties has been found, this does not mean that genome sequences always predict the potentially conditional expression of particular genes, nor their level of expression. Consequently, it will not always be possible to match phenotypic tests.

It is required that databases are public and that the data from food, (food-producing) animals, environment and humans are closely linked, in such a way that public health can profit the most from the strength of WGS. In Europe, both EFSA and ECDC have taken the initiative to encourage the collection of molecular data to ensure a better linkage of molecular typing data from humans to similar data from food and animals. A popular database is the Bacterial Isolate Genome Sequence database (BIGSdb) developed by Jolley and Maiden (2010). BIGSdb is able to handle NGS data of microbial genomes and can perform extended MLST.

The currently available databases suffer from several drawbacks. They lack harmonized protocols for submission, curation and storage of the data and interpretation, annotation and standardized data sharing formats are still missing.

10.3 Whole Genome Sequencing in Outbreak Investigation

Typing of the isolated organisms can aid in the detection of outbreak cases, can lead to a more specific case definition and is required to definitively link a source to the outbreak. Serotyping, PFGE, MLVA or any combination of these methods are currently used (PulseNet, www.cdc.gov/pulsenet/about/index.html). Serotyping does not have the resolution required for consistent detection of outbreaks. PFGE data is of fairly high resolution and is harmonized and shared in PulseNet, but it suffers from a certain lack in repeatability. In addition, for very monomorphic bacteria, like *Salmonella enterica* serovar Enteritidis, PFGE often does not have the required discriminatory power for outbreak investigation. As a result of the increased speed and decreased costs of acquisition and the high resolution of WGS, comparative genomic analysis of bacterial pathogens is rapidly becoming part of many outbreak investigations. However, the challenge for WGS is the large amount of data to be transferred and stored, as well as development of internationally harmonized analysis methods. While the development of tools for analysis and interpretation of WGS data is rapidly progressing, there is presently much uncertainty and certainly no consensus on how to translate these data into practically useful information for public health purposes. Nomenclature is, in its essence, a technique to reduce the amount of available information by assigning a short, yet still epidemiological informative code to isolates. For WGS data, one or more agreed nomenclature schemes are also required, since efficient communication between organizations is a prerequisite due to the international nature of infectious diseases. The currently used approach for nomenclature is under discussion. A nomenclature based on the earlier mentioned extended MLST approach is a good candidate, since it builds on the nomenclature of the already applied classical MLST. A prerequisite, however, is that all involved stakeholders need to agree on a specific scheme. Unique identifiers are given to unique alleles, which are subsequently curated in a global database. As an example of the potential of wgMLST for outbreak investigation, Schmid and colleagues showed in 2014 (Schmid 2014) that WGS is capable of discriminating *Listeria monocytogenes* SV1/2b clones that are not distinguishable by PFGE and fluorescent amplified fragment length polymorphism (fAFLP).

They investigated a cluster of seven human cases of listeriosis that occurred in Austria and in Germany between April 2011 and July 2013. In January 2013, the Austrian Food Authority mandated the Austrian Agency for Health and Food Safety to investigate the source and health related issues of an in 2012 dominant PFGE clone. Active case finding in Germany and Austria resulted in seven cases, all elderly women. Strains from food producers were also included in this study. The human and food related strains of *L. monocytogenes* SV1/2b could not be distinguished by PFGE and fAFLP. Epidemiological investigation, however, suggested that the Austrian cases were linked to one of the two Austrian food producers. To further analyse this listeriosis cluster, the seven human isolates, a control strain with a different PFGE/fAFLP profile and ten food isolates were subjected to whole genome sequencing (WGS). The investigators used MLST+, a whole genome wide MLST scheme based on 2298 genes that where present in all strains. MLST+ analysis suggested that the outbreak cases were linked with either one of the two Austrian (cheese or meat) food producers. It is worthwhile to mention that the products of the Austrian food producers appeared on the grocery bills collected from the outbreak cases. In this paper it was clearly shown that WGS was capable of discriminating *L. monocytogenes* SV1/2b clones that were not distinguishable by PFGE and fAFLP, hereby clearly showing the potential of WGS in outbreak investigations.

10.4 Viral Food-Borne Pathogens

The detection of viral food-borne illness relies on a combination of laboratory diagnosis, epidemiological investigation, pathogen typing and food trace-back investigations. There is a great challenge in reliable detection of viruses in food, a practice that is an essential part of outbreak investigations. Virus detection, in contrary to the detection of bacteria, cannot rely on a culture step. Food borne viruses are difficult to extract form food matrixes and the limit of detection is high. Another approach is the application of metagenomics based on deep sequencing, either by targeted sequencing of specific viruses or in an unbiased manner to determine the viral population. The detection and characterization of (novel) viruses are of paramount importance in outbreak investigation and the forecasting of future outbreaks of viral diseases in humans. Metagenomics studies based on deep sequencing in the analysis of clinical samples, natural reservoirs and food will help to control or prevent major outbreaks.

Due to the rapid rate of evolution of viruses, virologists have been using genome-wide sequencing for a long time. The interplay of epidemiological and evolutionary patterns, based on WGS, is emerging as a new field in public health microbiology. The availability of software such as BEAST (Drummond et al. 2012) has facilitated this strongly.

One important food borne virus is norovirus. Norovirus is a highly contagious particle which is easily transmitted via the faeces and vomit of infected people, but also through direct contact or through contaminated objects or via food and water. A network (NoroNet) of scientists working in public health institutes and universities share virological, epidemiological and molecular data on this virus. In order to type Noroviruses, several new tools and databases have been developed during the last couple of years. The database and tools are hosted by the RIVM (National Institute of Health and the Environment, The Netherlands).

10.5 Whole Genome Sequencing in Microbial Risk Assessment

The sequence data obtained by WGS contains all the genetic information of the strain. These data can be used for what is called 'comparative genomics', which is a term used for when the genetic content of two strains is compared. By doing this, short and long term evolutionary relationships and pathogenicity of food-borne pathogens, such as virulence, severity of disease, host specificity, ecological niche and mechanisms to adapt to particular (environmental) stress situations, can be determined. In addition, strain-, lineage- and niche-specific regions can be found andevents that resulted in the loss or acquisition of DNA can be identified. Comparison between the genome sequences of commensal (harmless) bacteria and those of pathogenic bacteria will help to identify genes that are involved in host specificity and to identify mechanisms of host-microorganism interaction.

As such, WGS has huge potential in the characterization of the genetic content of pathogens, in risk estimation of virulence potential, in antibiotic resistance profiles and in other health related properties. This information, in combination with phenotypic testing of bacterial behaviour (growth, survival, stress resistance, metabolism and adhesion to epithelial cells), provides a powerful combination to understand and control pathogens in the food supply chain. The outbreak strain causing the major outbreak in Germany in 2011 was characterized as an enteroaggregative *E. coli* (EAEC) that acquired toxin producing *stx*-genes by horizontal gene transfer. This resulted in a highly unusual combination of virulence traits, resulting in severe illness and death. The outbreak caused a paradigm change with respect to human pathogenicity of STEC and revealed substantial knowledge gaps in the emergence of hybrid types of pathogenic *E. coli.* A major obstacle for efficient monitoring, justifiable public health actions and efficient clinical management is the current inability to discriminate STEC strains posing a serious risk to humans (i.e. EHEC) from STEC strains that are not associated with severe and/or epidemic disease. This is primarily due to the huge variety in the presence/absence of virulence genes. Different classes of pathogenic strains are divided in five different seropathotypes (A–E), which is shown in Table 10.2. This model is based on the presence/absence of Shiga-toxin encoding genes (stx_1 and stx_2) and the intimin gene (*eae*) in relation to incidence in humans (reported frequency in human disease), association with outbreaks and the association with severe disease. To find

Table 10.2 Virulence markers in the seropathotype concept as proposed by Karmali et al. (2003)

Seropathotype	Incidence in human disease[a]	Outbreaks	Association with severe disease[b]	Virulence markers		Serotypes
				Vtx	Eae	
A	High	Common	Yes	Vtx2 (but may in addition also carry vtx1)	+	O157:H7, O157:NM
B	Moderate	Uncommon	Yes	Vtx1 and/or vtx2	+	O26:H11, O103:H2, O111:NM, O121:H19, O145NM
C	Low	Rare	Yes	Vtx1 and/or vtx2	±	O91:H21, O104:H21, O113:H21, O5:NM, O121:NM, O165:H25
D	Low	Rare	No	Vtx1 and/or vtx2	±	Multiple
E	Non-human only	NA[c]	NA[c]	Vtx1 and/or vtx2	±	Multiple

[a]Reported frequency in human disease
[b]Haemolytic uraemic syndrome (HUS) or haemorrhagic colitis (HC)
[c]NA = not applicable

out whether virulence genes or clusters of virulence genes could be linked to these different classes, the presence (or absence) of these genes was determined for a set of nearly 225 STEC strains that were isolated from humans, food and cattle by Franz and colleagues. The number of virulence genes in seropathotype A (with high incidence in humans and associated with severe disease, in which currently only STEC O157 and stx-producing EAEC O104:H4 are grouped) and seropathotype B (with moderate incidence in humans and associated with severe disease, in which among others O26 and O103 are grouped) were significantly higher than in STEC belonging to seropathotype C, D and E (Fig. 10.1). However, the situation was more complicated, as they also identified STEC serotypes with high numbers of virulence factors that were originally (based on epidemiological association) classified as seropathotype C (low incidence, associated with severe disease; O76:H19, O84:H-, O5:H-, O165:H-, O55:H7) and D (low incidence, not associated with severe disease; O6:H25, O80:H-, O101:H9). This shows that enforcement of food safety relying only on a predefined set of serotypes, those with a known historical epidemiological association with severe disease and outbreaks, is not sufficient and will not protect consumers from emerging pathogenic STEC.

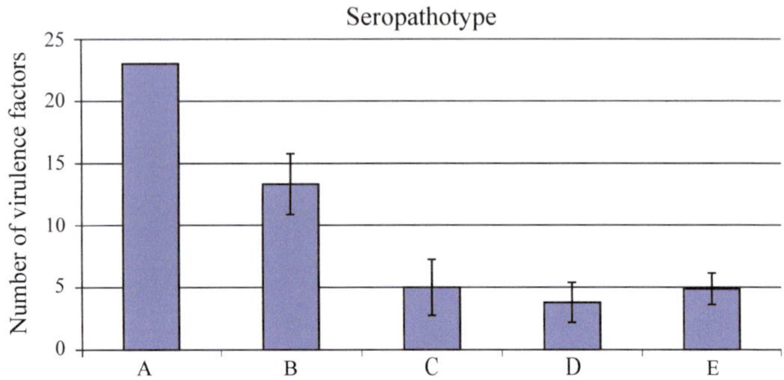

Fig. 10.1 Relation between the average numbers of virulence factors present in STEC isolates grouped in different seropathotypes (based on epidemiological association with severe disease and HUS)

In addition to PCR analysis, a subset of these strains was investigated by WGS and by studying the adherence to epithelial cells as an in vitro proxy for virulence. The sequences of the O157 strains were mapped to a reference genome (O157 Sakai) in order to identify SNPs. In total, 17 Single Nucleotide Polymorphisms (SNPs) could be identified that were significantly associated with an increased adherence to Caco-2 cells, a human gut epithelial cell line. These SNPs are shown in Table 10.3, together with the biological information about the genes involved (modified from Table 3 in Pielaat et al. 2015). As shown in this Table, 8 mutations lead to a change in the amino acid sequence of the protein and have a potential biological effect. Such mutations are called "nonsynonymous". The other 9 SNPs are called "synonymous" and do not alter the amino acid sequences. As an example, locus ECs1203 (antitermination protein Q encoded by prophage CP-933R) contains two SNPs (one synonymous and one non-synonymous) and is located upstream of the toxin producing gene (causing the severe disease haemolytic-uremic syndrome (HUS)) stx_2. Since the antitermination gene Q controls the level of Stx2 production and as it has been proposed that Stx2 promotes epithelial cell colonization, this gene is a potential candidate for an increased attachment marker. Maybe the most important finding of this study was that it revealed practical implications when using SNP data for risk assessment in a genome-wide association approach. These implications include optimum sample size for valid inference on population level, correction for population structure, quantification and calibration of results, reproducibility of the analysis, links with epidemiological data and linking to human health. Another major result of the study was the identification of genes with potential biological significance, although biological confirmation has not yet been performed. And as such, this study provides a first approach in (statistical) methodology development for WGS application in risk assessment.

In conclusion, the technical advance in sequencing whole genomes has enormous impact on the identification, typing and characterization of foodborne pathogens. WGS is becoming "the" method of choice for addressing foodborne outbreaks, attribution studies and for assessing the risk of food-borne pathogens.

Table 10.3 Biological information regarding the 17 significant SNPs, obtained using a model without correction for population structure; locus on the reference genome [position (bp)], function of this locus and description of the SNP (from Pielaat et al. 2015)

ID	Position (bp)	Locus tag Sakai	Function	SNP description[a]
1	808,227	ECs0729	RhsC protein	Synonymous; C219T
2	1,204,977	ECs1121	Prophage CP-933R tail fiber protein; putative host specificity protein	Synonymous; C1741T
3	1,265,758	ECs1203	Antitermination protein Q	Synonymous; C12T
4	1,265,760		Encoded by prophage CP-933R	Non-synonymous; G14A (R5Q)
5	1,955,401	ECs1977	Phage capsid and scaffold protein	Synonymous; C156T
6	1,963,016	ECs1987	Tail assembly protein	Synonymous; G351C/T
7	1,965,259	ECs1990	Prophage CP-933V tail fiber protein; putative host specificity protein	Synonymous; C1062T
8	2,168,378	ECs2164	Minor tail protein encoded by	Non-synonymous
9	2,168,379		Prophage CP-933O	T424G, C425A (S142E)
10	2,303,672	ECs2332	L-Arabinose 1-dehydrogenase	Non-synonymous; C268A (H90N)
11	3,115,509		Intergenic	G → A
12	3,480,394	ECs3489	Phage tail fiber protein encoded by prophage CP-933P	Synonymous; G252A
13	3,486,443	ECs3499	Hypothetical protein	Non-synonymous; T98C (L33S)
14	3,486,494			Non-synonymous; T149C (I50T)
15	4,929,010	ECs4864	RhsH protein	Non-synonymous; T134C (F45S)
16	5,054,140	ECs4969	Putative portal protein	Non-synonymous; G190A (E64K)
17	5,409,931	ECs5283	DNA-binding transcriptional repressor UxuR	Non-synonymous; C534A (N178K)

Note
[a]SNPs are displayed by type and position in the locus, followed in parentheses by the effect on the amino acid sequence in case of a non-synonymous SNP

10.6 Impact of WGS/NGS

1. WGS represents a huge step in diagnostic microbiological practice by giving a boost in the development of tools for detection, identification and characterization of food-borne pathogens

2. NGS allows identifying rare(r) variants that are otherwise missed due to the relative low costs to obtain high sequence coverage during a single run
3. NGS holds great promise for improving surveillance, (dispersed) outbreak investigation and the detection of emerging foodborne diseases, as genetic variation can be investigated at a much higher resolution, resulting in higher discriminatory power
4. NGS provides the opportunity to create a global system of linked databases for the identification and detailed genetic characterization of all microorganisms in clinical (and other) settings. Such a global system will strengthen local, national, and international surveillance of infectious diseases (Aarestrup et al. 2012)
5. Aarestrup et al. (2012) also discussed the advantage of the use of a single technology applied in different disciplines (e.g., bacteriology, virology, parasitology) and domains (human, food, animal, environment). This would facilitate global cross-cutting collaboration and information exchange (integrated surveillance), enabling rapid and coordinated responses to novel and known health threats
6. Sequencing of genomes will lead to a better identification of the phylogenetic relationships between strains and WGS is likely to become the method of choice for monitoring pathogens in time and geographically
7. Specific to food safety and food-borne illness, WGS has the potential capability to predict the virulence properties (and thereby the disease development), antimicrobial resistance (and thereby treatment options), host specificity (and thereby potential sources), etc. of an (outbreak) strain.

However, a successful introduction of Whole Genome Sequencing as the routine molecular (typing) method in outbreak investigation and in molecular risk assessment will need:

1. An agreed and interpretable epidemiological nomenclature for WGS typing
2. Agreed threshold levels for the degree of similarity that strains must comply with in order to make inference on whether an isolate is part of an outbreak or is unrelated to the outbreak. Over-discrimination might be a pitfall
3. As the volume of information contained in a genome sequence is vast, policies and security measures should be available to maintain the privacy and safety of this information.

References

Aarestrup FM, et al. Integrating Genome-based Informatics to Modernize Global Disease Monitoring, Information Sharing, and Response. EID. 2012;18(11).

Drummond AJ, et al. Bayesian phylogenetics with BEAUti and the BEAST 1.7. Mol Biol Evol. 2012;29:1969–73.

EFSA. Scientific Opinion on the evaluation of molecular typing methods for major food-borne microbiological hazards and their use for attribution modelling, outbreak investigation and scanning surveillance: Part 1 (evaluation of methods and applications. EFSA J. 2013;11:3502.

Franz E, et al. Genetic features differentiating bovine, food, and human isolates of Shiga toxin-producing *Escherichia coli* O157 in The Netherlands. J Clin Microbiol. 2012;50(3):772–80.

Foxman B, et al. Choosing an appropriate bacterial typing technique for epidemiologic studies. Epidemiol Perspect & Innov. 2005;2:10.

Jolley KA, Maiden MCJ. BIGSdb: scalable analysis of bacterial genome variation at the population level. BMC Bioinform. 2010;11:595–606.

Karmali MA, et al. Association of genomic O island 122 of Escherichia coli EDL 933 with verocytotoxin-producing *Escherichia coli* seropathotypes that are Linked to epidemic and/or serious disease. J Clin Microbiol. 2003;41(11):4930–40.

Pérez-Losada M, et al. Pathogen typing in the genomics era: MLST and the future of molecular epidemiology. Infect Genet Evol. 2013;3. http://dx.doi.org/10.1016/j.meegid.2013.01.009.4.

Pielaat A, et al. First step in using molecular data for microbial food safety risk assessment; hazard identification of *Escherichia coli* O157:H7 by coupling genomic data with in vitro adherence to human epithelial cells. Int J Food Microbiol. 2015. http://dx.doi.org/10.1016/j.ijfoodmicro.2015.04.009.

Sabat AJ, et al. Overview of molecular typing methods for outbreak detection and epidemiological surveillance (Review). Eurosurveillance. 2013;18(4).

Schmid D. Whole genome sequencing as a tool to investigate a cluster of seven cases of listeriosis in Austria and Germany, 2011–2013. Clin Microbiol Infect. 2014;20:431–6.

Recommended Literature

Laing C, Buchanan C, Taboada EN, Zhang Y, Kropinski A, Villegas A, Thomas JE, Gannon VPJ. Pan-genome sequence analysis using Panseq: an online tool for the rapid analysis of core and accessory genomic regions. BMC Bioinform. 2010;11.

Stasiewicz MJ, et al. Genomics tools in microbial food safety. Curr Opin Food Sci. 2015;4:105–10.

Strachan NJC, Rotariu O, Lopes B, Macrae M, Fairley S, Laing C, Gannon V, Allison LJ, Hanson MF, Dallman T, et al. Whole genome sequencing demonstrates that geographic variation of *Escherichia coli* O157 genotypes dominates host association. Sci Rep. 2015;5.

Struelens MJ, Brisse S. From molecular to genomic epidemiology: transforming surveillance and control of infectious diseases. Eurosurveillance. 2013;18(4).

Underwood AP, et al. Public health value of next-generation DNA sequencing of enterohemorrhagic *Escherichia coli* isolates from an outbreak. J Clin Microbiol. 2013;51:232–7.

Chapter 11
Veterinary Molecular Diagnostics

Hendrik I. Jan Roest, Marc Y. Engelsma, Eefke Weesendorp, Alex Bossers and Armin R. Elbers

Abstract In veterinary molecular diagnostics, samples originating from animals are tested. Developments in the farm animals sector and in our societal attitude towards pet animals have resulted in an increased demand for fast and reliable diagnostic techniques. Molecular diagnostics perfectly matches this increased demand. Veterinary molecular diagnostics primarily focuses on the detection, identification, and genotyping of pathogens. Techniques are comparable to those used in the human molecular diagnostic field. In veterinary diagnosis, these techniques are applied to either the diagnosis of diseases in individual animals and herds or to assess the disease status of a herd. Notable features of veterinary molecular diagnostics are the sampling of a diagnostic unit for herd diagnoses, which can compensate test characteristics and, to a lesser extent, the high RNA/DNA loads in cases of animal disease outbreaks. To further identify bacteria, numerous genotyping techniques are used, including whole genome sequencing (WGS). To characterise viruses, WGS is the method of choice. The applications of molecular techniques in molecular diagnostics and molecular epidemiology are presented in two case studies: Q fever and the highly pathogenic avian influenza.

Keywords Veterinary · Animal · Diagnostic unit · Herd diagnosis · Q-fever · Avian influenza

11.1 Introduction

In veterinary molecular diagnostics, samples are tested that originate from animals, which are generally collected by veterinarians. Farm animals like poultry, pigs, sheep, goats and cattle are kept on farms, while pet animals like dogs, cats, rodents, replies, birds, fish and so on, are mainly kept in household settings. In pet animal diagnosis, the individual animal is sampled and these samples are tested individually. In farm animal

H.I.J. Roest (✉) · M.Y. Engelsma · E. Weesendorp · A. Bossers · A.R. Elbers
Wageningen Bioveterinary Research, Lelystad, The Netherlands
e-mail: hendrikjan.roest@wur.nl

E. van Pelt-Verkuil et al. (eds.), *Molecular Diagnostics*,
DOI 10.1007/978-981-10-4511-0_11

diagnosis, however, the herd is the primary diagnostic unit, as these animals are kept in groups, flocks, and herds. To assess the status of the diagnostic unit, all animals can be sampled and tested, but other strategies can also be applied. Furthermore, although not unique to the veterinary field, very high loads of a pathogen can be present in a large number of animals in cases of outbreaks. This requires special attention during handling; processing and analyses of the samples, to avoid cross contamination. In this chapter, these aspects are discussed, a variety of molecular diagnostic applications in veterinary diagnostics is mentioned and two cases that cover the majority of veterinary molecular diagnostic techniques are discussed.

11.2 Background Information

Farm animal husbandry has changed in recent decades, with a tendency toward fewer farms with more animals per farm. The economic value of the sector has also increased. As a result of this, three main trends can be identified. First, with this increase in herd size, the health status of the animals has become more important. This has led to the implementation of monitoring programmes for endemic diseases, to carefully monitor and improve the health status of the herd. These programmes are based on diagnostic testing of the animals in the herd. Following on from this first trend, a second trend is the eradication of important notifiable animal diseases from the Netherlands. These diseases, like Foot and Mouth Disease, Classical Swine Fever, Highly Pathogenic Avian Influenza (HPAI), Tuberculosis and Brucellosis, had a devastating impact on the animals, making them useless for their economic purposes and a threat for public health. Due to the economic and public health impact of these diseases, there is a legal obligation to control them. In the Netherlands, stringent programmes have successfully eradicated most notifiable animal diseases in the past. To maintain and guarantee this disease free status, monitoring programmes based on diagnostic testing have been implemented. A third trend is the increase in animal trade. Large animals, as well as pet animals and horses, are transported all over the world. To minimise the risk of diseases spreading, the trade and transport of animals is highly regulated. Before they are transported to another country, the animals have to be tested, according to the demands of the receiving country, with tests that are prescribed by the World Organisation for Animal Health .

Public attitude towards pet animals has changed in recent decades as well. Pet animals have increasingly become a member of the family. This has changed the expectations from the veterinary medical sector. People go to specialised clinics for the best veterinary care and the pet animal veterinary profession has responded to this demand with a further specialisation. This specialisation has resulted in care that is comparable to human medical care, including an increase in the number and variety of diagnostic tests that are being performed.

Due to the increased monitoring and testing of animals, the need for fast and reliable diagnostic tools to detect pathogens has increased as well. Molecular diagnostics perfectly matches these increased demands. Veterinary molecular diagnostics primarily focuses on detection, identification and genotyping of pathogens. The techniques for molecular diagnosis of viral, bacterial and parasitic diseases of veterinary relevance are basically comparable to those used for the detection of human pathogens (see Vol. 1, Chaps. 1, 2, 3 and 4).

11.3 Veterinary Molecular Diagnostics

11.3.1 Veterinary Molecular Diagnostics in General

Molecular techniques are widely used in veterinary and livestock research for vaccine development, research on host-pathogens interactions, microbiome and resistome (the microbiome focussing on all antimicrobial resistance genes) analyses and the development of diagnostic assays and genome analysis to optimise breeding. In the veterinary diagnostic setting, molecular diagnostics primarily focuses on the detection, identification and genotyping of viruses, bacteria and parasites.

The different techniques and platforms used in the human medical diagnostic field are also available for veterinary applications. Real-time PCR platforms are most widely used, whereas in a few cases conventional PCR assays are still in use. Primer and probe design meet the basic principles described in this book. All optimisation of the PCR programmes, PCR mix and reagents follow these basic principles.

11.3.2 Individual Versus Herd Diagnostics

Pet animals are mainly kept in isolation or in small groups. Therefore, veterinary care and diagnostics for pet animals are comparable to human medicine. Molecular diagnostic tests focus on individuals and the assay must be able to detect pathogens with as high a degree of sensitivity as possible and, if clinically relevant, even detect very low levels of pathogens.

Farm animals, on the contrary, are kept in flocks or herds. In these groups, the interest shifts from the individual animal to the total number of animals in the group, this is called the diagnostic unit. Fences or bars separate the diagnostic units, which differ in the number of animals included. In sheep husbandry the diagnostic unit may vary from ten to over a hundred individual animals per herd. In poultry, a diagnostic unit may include as many as 100,000 individual birds.

To diagnose the presence or absence of disease or a pathogen in a diagnostic unit, all individual animals can be tested. However, with diagnostic units as large as

100,000 individual animals, testing all these individuals is very demanding and expensive and its efficacy is doubtful. So the question is: "does individual testing give more information when compared to other testing strategies" and "which margins for the test outcomes are acceptable?" This gives room for alternative testing strategies, which are based on a representative sample taken from the diagnostic unit. This sample must yield a reliable result, which is defined as a result that represents the status of the diagnostic unit with 100, 99, or 95% of confidence. Normally, a testing result that represents the status of the diagnostic unit with 99 or 95% confidence is accepted. It is presumed that the presence of a pathogen in a diagnostic unit has a normal distribution. A normal (or Gaussian) distribution is a continuous distribution with two parameters: mean and standard deviation. The normal distribution is symmetric around its mean (see Vol. 1, Chap. 6). Then, given the assumed prevalence expected when the disease agent would have infected a herd or flock, the analytical sensitivity and specificity of a test, and the required confidence in our estimate, the required size of the sample of the diagnostic unit can be calculated. This sample size is a representation of the diagnostic unit. It is important to note that the sample size calculation also includes the test characteristics. An assay with test characteristics that are suboptimal for individual diagnostic testing (as used in pet animal and human diagnostic testing) can still be useful in herd testing. This gives more flexibility in the assays that can be used. For example, when an assay has an analytical sensitivity of less than 100% on an individual test basis, this can be compensated for on a herd basis by increasing the samples size:

$$\text{hSe (herd sensitivity)} = 1 - (1 - [\text{Se} * \text{P} + (1 - \text{Sp}) * (1 - \text{P})])^{\text{n}}$$

with P = Prevalence (fraction between 0 and 1), and Se = analytical sensitivity (fraction between 0 and 1) and Sp = analytical specificity (fraction between 0 and 1) and n = number of animals sampled.

The consequence of increasing the herd sensitivity (hSe) by increasing the sample size is that the herd specificity (hSp) decreases. Therefore positive samples have to be retested in another diagnostic test to identify the false-positive samples.

$$\text{hSp} = (\text{Sp})^{\text{n}}$$

With Sp = analytical specificity and n = number of animals sampled.

Example I, the sample size can compensate for a lower sensitivity

We use an assay with Se = 0.90 and Sp = 0.995 to diagnose the presence of a certain disease on a dairy farm, with an expected prevalence of 20%. If we would sample 36 animals within the group of dairy cows, the calculated hSe would be: $1 - (1 - [0.90 * 0.2 + (1 - 0.995) * (1 - 0.2)])^{36} = 0.9993$ (so almost optimal).

However when we now use a test with an even lower Se, like Se = 0.60 and Sp = 0.995 and when we want to have the same hSe of 0.9993 we have to sample: $0.9993 = 1 - (1 - [0.60 * 0.2 + (1 - 0.995) * (1 - 0.2)])^{n} = 1 - 0.876^{n}$. From this we get $0.9993 - 1 = -0.876^n$, and so $0.0007 = 0.876^n$. Taking the Log on both sides: $\text{Log}(0.0007) = n * \text{Log}(0.876)$ and $n = \text{Log}(0.0007)/\text{Log}(0.876) = 55$ animals.

So taking a larger sample will compensate for the poor analytical sensitivity of the test.

Example II, the need for confirmation of positive results

We use an assay with Se = 0.90 and Sp = 0.995 to diagnose the presence of a certain disease on a dairy farm, with an expected prevalence of 20%. If we would sample 36 animals within the group of dairy cows, the calculated hSe would be: $1 - (1 - [0.90 * 0.2 + (1 - 0.995) * (1 - 0.2)])^{36} = 0.9993$ (so almost optimal); however, the calculated $hSp = (0.995)^{36} = 0.835$, so with testing the 36 animals, we expect false positive test results (in this case 16.5%); the positive tests have to be confirmed in another diagnostic test with a better specificity.

In summary, both individual diagnostics and herd diagnostics are applied in veterinary molecular diagnostics. In herd diagnostics, the diagnostic unit is important and a sample size that represents the diagnostic unit and takes the test characteristics into account can be calculated.

11.3.3 High Loads of Pathogens

A major concern in veterinary molecular diagnostics is the sometimes high viral or bacterial loads in diagnostic samples. These can result in highly concentrated RNA or DNA samples after the isolation procedure. This is especially the case when animals are infected with highly infectious viruses that cause, for example, Food and Mouth Disease or Classical Swine Fever or with bacteria like *Coxiella burnetii* (the bacterium that causes Q fever). The Q fever bacterium can cause abortion in pregnant goats. Samples of infected placentas can contain very high amounts of pathogens. qPCR tests of the undiluted sample can result in Cq values below 10. These highly concentrated RNA/DNA samples can lead to cross contamination during sampling, handling and processing (e.g. while pipetting) of samples or lead to contamination of equipment or laboratories. The generic setup of the routine PCR laboratory and the measures used should be strictly implemented (see Vol. 1, Chap. 6). These measures include the use of dedicated PCR rooms for the different steps in the PCR, sample handling in dedicated hoods, and the use of negative controls (Pestana et al. 2010).

11.3.4 Applications of Veterinary Molecular Diagnostics

As mentioned already, individual diagnostics and herd diagnostics can be distinguished in veterinary laboratory diagnosis. In the individual and herd diagnosis of infectious diseases, the primary goal is to diagnose the presence or absence of a pathogen that might cause disease. A correct diagnosis is the basis of an action plan or treatment plan to treat the animals or to avoid disease in the future, by taking vaccination or management measures, for example. Herd diagnostics can be used in two ways: for diagnostic purposes and for monitoring the specific disease status of a herd. A special application of veterinary molecular diagnostics is the diagnosis of notifiable animal diseases. The consequence of a positive test result might be the culling of the herd.

Individual veterinary molecular diagnostics are applied in a similar way to human molecular diagnostics. The diagnostic assay focuses on animals that are ill and specific tests can be performed.

In herd diagnosis of infectious diseases, the diagnostic unit is the group of affected animals, so a sample of the animals with clinical symptoms representative for the disease is needed. For example, when calves are coughing, symptomatic animals must be sampled. These samples can be tested for several viruses and bacteria. qPCR assays are available to detect viruses like Bovine Viral Diarrhoea virus, Bovine Respiratory Syncytial virus, Parainfluenza virus and Bovine Herpes virus 1. Also bacteria like *Mannheimia haemolytica* and various *Mycoplasma* species can be detected by qPCR assays. As all these agents can cause or play a role in respiratory problems in calves, it would be ideal if these agents could be detected in a multiplex real-time PCR. Research on the composition of such a multiplex PCR is ongoing and focuses on the uniform set up of the PCR for the different pathogens, so that these PCRs can be integrated into one multiplex assay.

The health status of a herd for a specific disease (determined for example as part of a monitoring programme or to assess the status prior to the start of an eradication or monitoring programme) should be assessed by testing a representative sample of the diagnostic unit. In a monitoring programme, this is done at certain time points. For most diseases this is done by serological assays, but in some cases this is not possible. An example of this is the monitoring of the Q fever status of dairy goat farms (see the paragraph on 'The Dutch Q fever outbreak' in this chapter). As mandatory vaccination interferes with serological tests, qPCR is the method of choice. In this specific case, the sample of the diagnostic unit is a bulk tank milk sample, on which the qPCR is performed. Validation of this method revealed that this system is sensitive enough to detect only a few Q fever-bacteria shedding animals in a dairy goat herd of thousands of animals.

For the diagnosis of notifiable animal diseases, techniques are shifting from conventional techniques, like virus and bacteria isolation methods, to qPCR. This is a relatively slow process, as the diagnostic assays for notifiable diseases are regulated in legislation and it is important that the laboratory diagnosis is absolutely reliable, due to the considerable consequences of a positive result. For most notifiable diseases, a positive result will mean culling the animals and implementing additional control measures, like a transport ban, to avoid further spread of the disease.

In summary, molecular techniques in veterinary medicine are used at the individual and herd levels for diagnostic and monitoring purposes.

11.3.5 Detection, Typing and Whole Genome Sequencing

The outcome of a PCR is the presence or absence of DNA or RNA of a specific pathogen in a sample. For the detection of a specific pathogen in a sample this is sufficient. In certain cases additional information, for example on the subspecies of the pathogen, is needed for the diagnosis or for the treatment plan. This additional information can be obtained with genotyping techniques, like multi-locus sequence typing (MLST), multi-locus variable number tandem repeat analysis (MLVA), single nucleotide polymorphism (SNP) typing or 16S rDNA sequencing. With genotyping, more genetic information of the pathogen is incorporated in the test, revealing a better characterisation. These techniques are applied to pathogens with genome sizes comparable to bacteria (genome size range: 130 kbp to >14 Mbp) rather than on viruses (genome size range: 2 kbp–2 Mbp). Sometimes it is needed to take into account all genome information; this can be done with whole genome sequencing (WGS). For viruses, with their smaller genomes, WGS is easier to perform than for bacteria. An example for the application of WGS of viruses is presented in the section 'Highly pathogenic avian influenza H5N8 epidemic in the Netherlands in 2014', but for bacteria WGS is applicable as well (Maio et al. 2014).

For genotyping techniques, it is important to know how well the technique can distinguish unrelated bacterial strains. This characteristic of genotyping techniques is called the discriminatory power. This is determined by the number of genotypes defined by the genotyping method and their relative frequencies. Hunter and Gaston suggest a single numerical index of discrimination. The Hunter-Gaston Diversity Index (HGDI) is based on the probability that two unrelated strains sampled from a test population will be placed into different typing groups (Hunter et al. 1988). By comparing the HGDI of a typing system, the discriminatory power can be evaluated. It is important to note that the calculated HGDI depends on the panel of strains (i.e. relatedness of the strains that are used to evaluate the discriminatory power of a genotyping method). For an unbiased comparison of typing methods the same panel should be used. Despite its importance for the quality of typing systems, the discriminatory power is not assessed for most of the typing systems available. However, for *C. burnetii*, the agent causing Q fever, the discriminatory power is calculated for several genotyping systems that are described in literature. Restriction fragment length polymorphism (RFLP)-typing has a HGDI of 0.86. The two MLVA typing panels that have been described have a HGDI of 0.92. When these two panels are combined, the HGDI increases to 0.99. It is suggested that a HGDI of >0.90 is desirable to interpret typing results with confidence, indicating the MLVA typing system as a useful typing tool for *C. burnetii*. Combining the two panels of the MLVA genotyping system will result in a better discrimination between strains than using one of the panels alone.

For zoonotic bacteria (i.e. bacteria that can be transmitted from animals to humans) it is important that the genotyping systems used for the same organism in the human and veterinary field are harmonised, so that the results can be compared. This allows possible transmission from animals to humans to be assessed and the possible animal source of human disease to be identified. Harmonised genotyping systems are also required for molecular epidemiology of zoonotic diseases and antimicrobial resistance. This knowledge is essential to improve control and prevention. An example of this is given in the section '*Molecular aspects of the Dutch Q fever outbreak*' in this chapter.

In conclusion, molecular techniques, from detection to WGS, are applied in the veterinary field to detect specific agents or to reveal as much genetic information as is possible to identify the agent.

11.4 Molecular Aspects of the Dutch Q Fever Outbreak

11.4.1 Introduction

Q fever is a disease of humans and animals, which is caused by the intracellular bacterium *C. burnetii*. Q fever has a worldwide distribution, except for New Zealand. Although many animal species can be infected by *C. burnetii*, the primary sources of *C. burnetii* for humans are small ruminants. When *C. burnetii* infects pregnant small ruminants, the bacterium invades the placenta and multiplies there to up to 10^9 organisms per gram of placenta. During abortion (early birth) or normal birth, these organisms are excreted into the environment and can infect humans via direct contact or via aerosol transmission. Clinical symptoms in humans vary from mild flu-like symptoms to pneumonia and hepatitis, with aneurism of the aorta and inflammation of heart valves observed in chronic cases.

11.4.2 The Dutch Q Fever Outbreak

Since the 1950s Q fever has been present in the Netherlands. However, only a few human cases were reported until the 1970s. In 1975, Q fever became notifiable for humans and since that time about 20 cases were reported on average each year. The first human outbreak of Q fever in the Netherlands was reported in 2007: 168 cases were registered in the south of the Netherlands. In 2008 and 2009 the annual number of human cases increased to 1000 and 2355, respectively. Over the years 2007–2010, 4000 human cases were reported, making the Dutch Q fever outbreak the largest laboratory confirmed Q fever outbreak ever seen.

Until 2005, serological testing revealed that Q fever was present in the animal population, although no clinical symptoms were described. This changed in 2005. Slightly ahead of the human outbreak, Q fever problems started in the dairy goat and dairy sheep population with abortion rates of up to 80% per herd. Between 2005 and 2009, significant abortions were registered on 28 dairy goat farms and 2 dairy sheep farms. With goat herd sizes of between 600 and 7000 animals, huge amounts of *C. burnetii* were spread into the environment during abortion and early birth. These bacteria were transported to the neighbouring human population by the prevailing northeast winds during quite dry spring periods. All this took place in the southeast part of the Netherlands, which is highly populated and has a dense dairy goat industry. As goats were the suspects of the human Q fever outbreaks, increasingly strong measures were imposed to prevent the spreading of *C. burnetii* in the lambing season. The first measures were implemented in 2008 and included the notification of abortions at small ruminant holdings and hygiene measures, especially on manure handling and a voluntary vaccination. In 2009, the measures were extended with stricter notification criteria, including positivity of the bulk tank milk for *C. burnetii* DNA, a transport and breeding ban, and a mandatory vaccination. All these measures, however, did not prevent the increase of human cases in 2009. To ultimately stop the increase of human cases, the drastic decision was taken in 2010 to eliminate all potential high-risk animals. This resulted in the culling of all pregnant goats on Q fever positive farms. All these measures finally resulted in a decrease in the annual number of human cases in 2010. This decline was consistent over the following years and by the end of 2012 the Q fever outbreak was declared to have ended.

11.4.3 Molecular Aspects of the Dutch Q Fever Outbreak

Q fever became a notifiable animal disease in 2008, as mentioned above. Goats that were suspected of abortion due to Q fever were sampled via a vaginal swab. At the national reference laboratory for animal diseases, the Wageningen Bioveterinary Research (WBVR) this vaginal swab was tested using a qPCR targeting the IS*1111* element in the *C. burnetii* genome. This IS*1111* element is specific for *C. burnetii* and present in multiple copies in the genome of the bacterium. Therefore the PCR targeting of this element is more sensitive when compared to qPCRs that are targeting single-copy genes, like Com*1*. With the stricter notification criteria in 2009, bulk tank milk was also being tested. For this, a testing scheme was constructed in which the Animal Health Service (AHS) did the initial screening with a IS*1111* targeting qPCR and this was confirmed at WBVR. The primers used in both qPCRs used at the AHS and at WBVR were slightly different, so the WBVR qPCR could be used as a confirmation test. With this bulk tank milk sample, the herd, which is the diagnostic unit, was sampled and tested.

Eventually, the connection between dairy goats and humans was primary based on epidemiological findings. Spatial distribution and the succession in time between the Q fever outbreaks in dairy goat herds and in humans pointed towards a relation. This relation was confirmed by genotyping. First the genomes of the Q fever bacteria that caused the various abortions in dairy goat herds were compared by a multi-locus variable number tandem repeat analysis (MLVA) technique (see Vol. 1, Chap. 5). The number of microsatellite repeats (see Vol. 1, Chap. 2) at specific areas of the genome were counted and compared between the different strains.

This MLVA genotyping method appeared to have a good discriminatory power with a HGDI of 0.99. For the goats, it became clear that one genotype was predominantly present at all dairy goat farms with abortion problems (Fig. 11.1). The same was done for a large number of human Q fever cases and results revealed that two very closely related genotypes were predominating (Fig. 11.2). Comparison between the predominant genotypes in both dairy goats and humans showed up to 100% agreement (Table 11.1). This confirmed that *C. burnetii* excreted by the dairy goats during abortions caused the Q fever outbreak in humans.

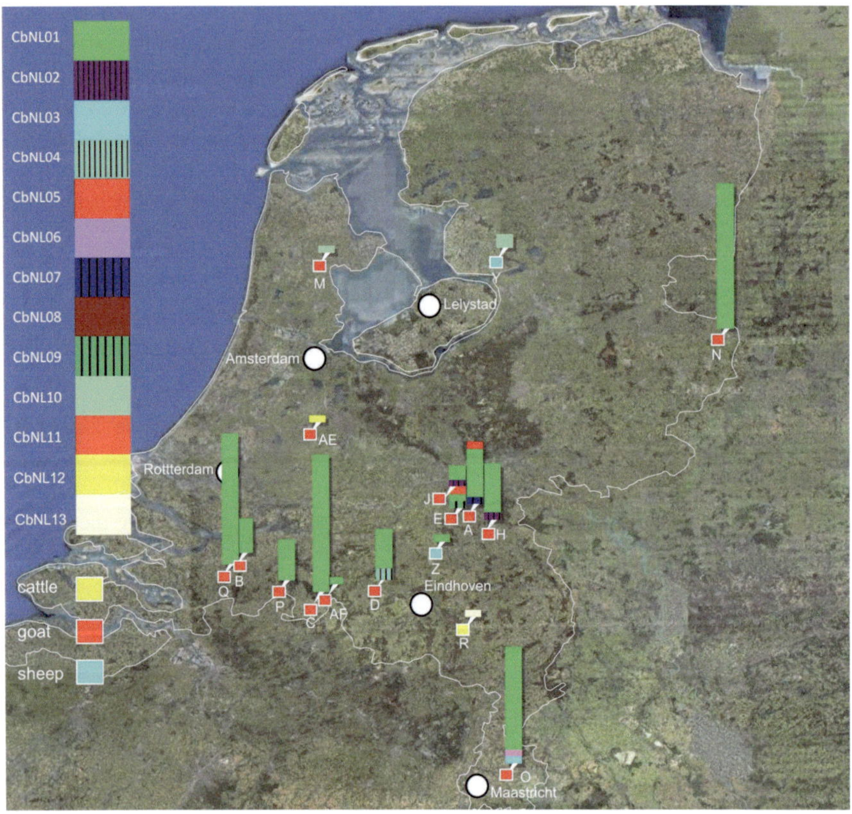

Fig. 11.1 Map of the Netherlands showing locations of farms sampled during the Q fever outbreak, 2007–2010. Farms (*squares*) are indicated by *letter*, the animal species present at the farm is indicated by *colour*; *bars* at each farm's location indicate genotypes of *Coxiella burnetii* found per farm. The *height of the bar* indicates numbers of isolates per genotype (*source* Roest et al. 2011)

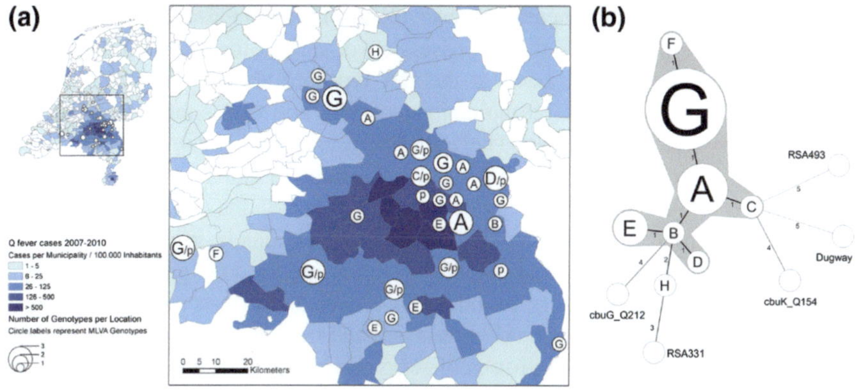

Fig. 11.2 Geographical distribution and relationship between Dutch *C. burnetii* genotypes from humans. **a** Geographical locations of the MLVA genotypes obtained from the 2007–2010 outbreak. **b** Minimum spanning tree showing the relationship between the obtained MLVA genotypes identified (*source* Tilburg et al. 2012)

Table 11.1 Combination of the *C. burnetii* genotyping results from humans and goats as presented in Figs. 11.1 and 11.2

	Ms						Prevalence (%)
	23	24	27	28	33	34	
Humans A	6	11	3	3	2	7	21
Humans G	3	11	3	3	2	7	55
Goat, CbNL01	3	11	3	3	2	7	90

Presented are the microsatellite (Ms) loci that are overlapping in the two multi-locus variable number tandem repeat analysis (MLVA) used in the studies. Results show that the number of repeats for the different Ms in both studies are similar, except for Ms 23 in genotype A, meaning that the strains of *C. burnetii* from goats and humans are very closely related or similar

11.5 Highly Pathogenic Avian Influenza H5N8 Epidemic Outbreak in the Netherlands in 2014

11.5.1 Introduction

Avian Influenza (AI) is a viral disease of birds, which is caused by influenza A viruses. Influenza A viruses that infect poultry can be divided into two groups: low-pathogenic (LPAI) and highly-pathogenic avian influenza (HPAI) viruses on the basis of severity of the disease following experimental infection of chickens. Both LPAI and HPAI are notifiable diseases in poultry. Waterfowl and shorebirds (wild and domestic) form the major natural reservoir and source of all known influenza A viruses. The virus particles carry two glycoproteins on their surface:

haemagglutinin (HA) and neuraminidase (NA). The influenza A virus can be divided into subtypes, based on the possession of one of the 16 distinct haemagglutinin antigens (H1–H16). Each of the HA types combines with one of 9 NA types (N1–N9). Virtually all combinations of HA and NA subtypes have been isolated from wild bird species. The HPAI viruses that cause major diseases in poultry, belong to the H5 and H7 subtypes, although not all viruses of these subtypes are HPAI viruses. The influenza A viruses of the remaining subtypes belong to the group of LPAI viruses. HPAI is an extremely infectious and fatal poultry disease. However, most AI virus strains are of low pathogenicity and typically cause mild respiratory problems or a decrease in egg production and/or water and feed intake. Nearly all AI viruses isolated from wild birds are LPAI. Only after the introduction of LPAI viruses of subtype H5 or H7 in poultry, particularly in chickens and turkeys, can HPAI mutants emerge after a variable length of time. However, in 2005, thousands of wild waterfowl were found dead at Qinghai Lake, in central China. HPAI-H5N1 was detected in wild birds, some of which were migratory. This event was a distinct turning point in the spread of the virus. By early 2006, it had been detected across large parts of South Asia, Western Europe, and parts of Africa, spread by wild birds.

11.5.2 Background of the Epidemic

The highly pathogenic avian influenza (HPAI) virus of subtype H5N8 probably originated in China, where it had already been isolated in 2009–2010. Pathogenicity studies showed that the virus was highly virulent in chickens but mildly or moderately virulent in wild ducks. Phylogenetic research demonstrated that it was the product of various gene re-assortment events. The RNA of the virus (influenza viruses are RNA viruses) consists of segments that come from several different influenza viruses. The backbone of the HPAI H5N8 virus is formed by parts of the HPAI H5N1 virus that has circulated in China since 1997 and spread worldwide since 2004. Beginning in January 2014, a HPAI virus of subtype H5N8 spread rapidly in South Korea, initially among farmed ducks. During the first outbreaks among farmed ducks, numerous dead Baikal teals (*Anas formosa*, a species of migratory wild ducks) were found near the affected duck farms, leading to the hypothesis that the wild ducks carried the infection. Genetic analysis of virus isolates from infected domesticated ducks and dead Baikal teals in the surrounding area in South Korea indicated that these isolates strongly resembled earlier Chinese isolates. The analysis also revealed that the HPAI H5N8 virus in South Korea is a product of re-assortment of A/duck/Jiangsu/k1203/2010 (H5N8) and other avian influenza viruses that co-circulated in birds in East Asia during 2009–2012.

Kang et al. (2015) demonstrated by experimental infection of wild ducks (*Anas platyrhynchos*) and Baikal teals that HPAI H5N1 and H5N8 virus isolates did not cause serious illness or death in these birds. Recent phylogenetic studies of HPAI H5H8 viruses isolated from infected poultry and wild birds in 2014 in South Korea showed that migrating birds played a key role in the introduction and spread of the virus in the initial phase of the 2014 epidemic. In mid-April 2014, the presence of HPAI H5N8 virus was demonstrated at a poultry farm in Japan, after a rise in the death rate was noted. During a monitoring programme in November 2014 in Japan, faecal samples of migrating Bewick's Tundra swans (*Cygnus columbianus bewickii*) tested positive for the HPAI H5N8 virus.

In the weekend ofthe first and second of November 2014, increased mortality was observed on a turkey farm with approximately 35,000 turkeys in Heinrichswalde, Mecklenburg-West Pomerania in Germany. HPAI H5N8 virus was confirmed on the fifth of November, after avian flu experts at a state veterinary research organisation, the Friedrich Loeffler Institute, carried out diagnostic tests. Mecklenburg-West Pomerania, in northeast Germany, borders the Baltic Sea to the north, and Poland to the east. The state has more than 1700 lakes. The source of infection of the turkey farm was most likely wild birds.

11.5.3 First HPAI H5N8 Virus Outbreak in the Netherlands: In Search for the Source of the Virus

Just a few days after confirmation of a HPAI H5N8 outbreak in turkeys in Germany, laying chickens in one of six poultry houses on an indoor-layer farm with 124,000 hens in the western part of the Netherlands began dying at an exponentially increasing rate. The dead chickens were submitted for necropsy to the Dutch Animal Health Service on the 14th of November, to investigate the reason for this increased mortality. As a standard procedure, because a high mortality rate is also one of the signs of the notifiable disease HPAI, the swab samples were forwarded to CVI, the Netherlands' national reference laboratory for animal diseases. There, RNA was extracted from cloacal and oropharyngeal samples of clinically affected hens with positive results from the screening influenza RT-qPCR, which detects all avian influenza virus subtypes. Positive screening samples were checked for the presence of H5 and H7 influenza subtypes by RT-qPCR as recommended by the European Union reference laboratory. Hemagglutinin (HA) and neuraminidase (NA) sequence analysis was performed on the PCR fragments. The HA cleavage site showed polybasic properties RNSPLRERRRKR*GLFGAIA, confirming the high pathogenicity of the virus. In addition, HA and NA sequence results showed that the virus subtype was H5N8. Preliminary sequencing of the cleavage site showed that it shared high similarity with that of the outbreak strain from Germany. However, full-length sequencing was necessary to elucidate the origin of the HPAI (H5N8) virus detected in the Netherlands. All eight RNA genome segments of the outbreak virus were amplified. Purified amplicons were sequenced at high coverage (average > 1000), generating paired-end read lengths of 150 bases. High-quality, quality-control–passed sequence reads were iteratively mapped against the genome sequence of the H5N8 virus from South Korea (GenBank accession numbers KJ511809–KJ511816) to generate a majority (>80% evidence) consensus sequence of all segments. The consensus sequences were compared with de novo–assembled sequence reads and substantial differences were not detected. A molecular phylogenetic analysis on all nucleic acid sequences was subsequently performed by using the maximum-likelihood method (see Vol. 1, Chaps. 8 and 9). Genetic analysis showed that the H5N8 virus from the Netherlands (A/chicken/Netherlands/14015526/2014) and viruses from poultry and wild birds from Europe and two strains from Japan (A/duck/Chiba/26-372-48/2014 and A/duck/Chiba/26-372-61/2014) which were detected thereafter, were closely related. Analysis also showed that the viruses are descen-dants of three strains isolated in early 2014: A/broiler duck/Buan2/2014 and A/Baikal teal/Korea/Donglim3/2014 from South Korea and A/chicken/Kumamoto/1-7/2014 from Japan. HPAI H5N8 virus isolated from a wigeon (*Anas penelope*) in Sakha, northeast Russia, is a precursor, phy-logenetically located at the node of European and Chiba viruses. This virus was isolated in September 2014, but the sequence was only released in December 2014. The number of per site base substitutions between the sequences was determined using maximum composite likelihood. Sequences of the eight genome segments in the H5N8 virus from the Netherlands differed from those of strains A/broiler duck/Korea/Buan2/2014, A/Baikal teal/Korea/Donglim3/2014, and A/chicken/Kumamoto/1-7/2014 by a minimum of 0 and a maximum of 0.009 substitutions. On the basis of data in the National Centre for Biotechnology Information Influenza Virus Resource (http://www.ncbi.nlm.nih.gov/genomes/FLU/FLU.html) and EpiFlu (http://www.gisaid.org) databases, the H5N8 virus from the Netherlands shared the highest similarity with strain A/Baikal teal/Korea/Donglim3/2014 from South Korea.

11.5.4 All Five Outbreaks of HPAI H5N8 Virus in the Netherlands: Search for Transmission Routes

After the first outbreak, four more poultry farms became infected. In the last one, infection was confirmed at the end of November, following the observation of exponentially increasing mortality. Backward and forward tracing of possible dangerous contacts in the framework of the standard epidemiological investigation (i.e. transport, professional visitors such as advisors, veterinary practitioner, etc.; possibly contaminated materials delivered to the farm such as feed, bedding etc.; possibly contaminated transport vehicles, etc.) by the Dutch Food and Safety Authority revealed no indication for dangerous contacts of such connections between HPAI H5N8 outbreaks in Asia, Germany, the United Kingdom and the Netherlands between October and December 2014. Moreover, no contacts were discovered between the outbreaks in the Netherlands. The question arose as to whether the outbreaks on the five Dutch poultry farms were caused by separate virus introductions or by transmission between farms. To answer this question, the complete virus RNA genome obtained from several animals from each farm was sequenced. The aim of this research was to assess possible routes of transmission of the virus by sequence and temporal phylogenetic analysis. The five fully sequenced Dutch H5N8 sequences were each aligned with 22 H5N8 sequences obtained from GISAID. Phylogenetic analysis of highly pathogenic avian influenza A(H5N8) virus strains causing outbreaks in Dutch poultry farms in 2014 provided evidence for separate introduction of the virus in four outbreaks in farms located 16–112 km from each other and for one between-farm transmission between the third and fourth outbreak in farms located 550 m from each other (Fig. 11.3). The findings also suggested that the Dutch outbreak virus strain 'Ter Aar' (the second outbreak in the Netherlands) and the first German outbreak strain from 2014 shared a common ancestor. In addition, the data indicated that the Dutch outbreak viruses descended from a H5N8 virus that circulated in Asia around 2009, possibly China, and subsequently spread to South Korea and Japan and finally to Europe as well. Evolution of the virus seemed to follow a parallel track in Japan and Europe, which supports the hypothesis that H5N8 virus was exchanged between migratory wild waterfowl at their breeding grounds in Siberia and from there was carried by migrating waterfowl to Europe.

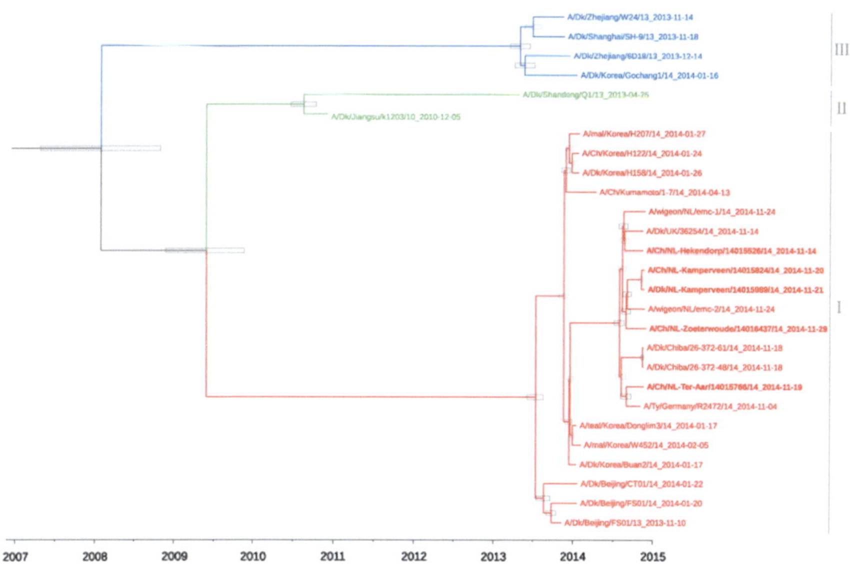

Fig. 11.3 Time-scaled phylogenies of the sequences of the viruses involved in the Dutch highly pathogenic avian influenza (HPAI) H5N8 outbreaks (*in bold*) with its closest relatives (*source* Bouwstra et al. 2015)

References

Bouwstra RJ, et al. Full genome sequence of HPAI H5N8 outbreak strains provide evidence for four separate introductions and one between-poultry farm transmission in the Netherlands, 2014. EuroSurveillance. 2015;20(26):21174. doi:10.2807/1560-7917.ES2015.20.26.21174.

Hunter PR, et al. Numerical index of the discriminatory ability of typing systems: an application of Simpson's index of diversity. J Clin Microbiol. 1988;26:2465–6.

Kang HM, et al. Novel reassortant influenza A(H5N8) viruses among inoculated domestic and wild ducks, South Korea, 2014. Emerg Infect Dis. 2015;21:298–304. doi:10.3201/eid2102.141268.

Maio E, et al. Identification and typing of *Brucella* spp. in stranded harbour porpoises (*Phocoena phocoena*) on the Dutch coast. Vet Microbiol. 2014;173(1–2):118–24. doi:10.1016/j.vetmic.2014.07.010.

Pestana EA, et al. PCR laboratory set-up. In: Rapid and sensitive veterinary molecular diagnostics—real time PCR applications. Dordrecht: Springer; 2010. p. 235–246. http://dx.doi.org/10.1007/978-90-481-3132-7_6.

Roest HIJ, et al. Molecular epidemiology of *Coxiella burnetii* from ruminants in Q fever outbreak, the Netherlands. Emerg Infect Dis. 2011;17(4):668–75. doi:10.3201/eid1704.101562.

Tilburg JJ, et al. Genotypic diversity of *Coxiella burnetii* in the 2007–2010 Q fever outbreak episodes in The Netherlands. J Clin Microbiol. 2012;50(3):1076–8. doi:10.1128/JCM.05497-11.

Index

E. van Pelt-Verkuil et al. (eds.), *Molecular Diagnostics*,
DOI 10.1007/978-981-10-4511-0

FSC
www.fsc.org
MIX
Papier aus verantwortungsvollen Quellen
Paper from responsible sources
FSC® C105338

If you have any concerns about our products, you can contact us on
ProductSafety@springernature.com

In case Publisher is established outside the EU, the EU authorized representative is:
Springer Nature Customer Service Center GmbH
Europaplatz 3, 69115 Heidelberg, Germany

Printed by Libri Plureos GmbH
in Hamburg, Germany